Emma Domínguez-Rué, Linda Nierling (eds.)
Ageing and Technology

Emma Domínguez-Rué, Linda Nierling (eds.)
Ageing and Technology
Perspectives from the Social Sciences

[transcript]

Universitat de Lleida   Karlsruhe Institute of Technology

Bibliographic information published by the Deutsche Nationalbibliothek
The Deutsche Nationalbibliothek lists this publication in the Deutsche Nationalbibliografie; detailed bibliographic data are available in the Internet at http://dnb.d-nb.de

© 2016 transcript Verlag, Bielefeld

All rights reserved. No part of this book may be reprinted or reproduced or utilized in any form or by any electronic, mechanical, or other means, now known or hereafter invented, including photocopying and recording, or in any information storage or retrieval system, without permission in writing from the publisher.

Cover layout: Kordula Röckenhaus, Bielefeld
Printed in Germany
Print-ISBN 978-3-8376-2957-6
PDF-ISBN 978-3-8394-2957-0

TABLE OF CONTENTS

All that Glitters is not Silver – Technologies for the Elderly in Context.
Introduction
Linda Nierling and Emma Domínguez-Rué | 9

AGEING, TECHNOLOGY AND (INTER-)PERSONAL
DEVELOPMENT: OLD AGERS AS TECHNOLOGY USERS

Motives of the Elderly for the Use of Technology in their Daily Lives
Helga Pelizäus-Hoffmeister | 27

An Exploration of Mobile Telephony Non-use among Older People
Mireia Fernández-Ardèvol | 47

Older Women on the Game: Understanding Digital Game
Perspectives from an Ageing Cohort
Hannah Marston and Sheri Graner-Ray | 67

Social Inclusion of Elderly People in Rural Areas by Social and
Technological Mechanisms
Peter Biniok, Iris Menke, Stefan Selke | 93

AGEING, TECHNOLOGY AND ELDERLY CARE: ASSISTIVE
TECHNOLOGIES

Skripting Age – The Negotiation of Age and Aging in Ambient
Assisted Living
Cordula Endter | 121

Making Space for Ageing: Embedding Social and Psychological Needs of Older People into Smart Home Technology
Barry Guihen | 141

Seeing Again: Dementia, Personhood and Technology
Ike Kamphof | 163

Enabling a Mobile and Independent Way of Life for People with Dementia – Needs-oriented Technology Development
Nora Weinberger, Bettina-Johanna Krings, Michael Decker | 183

Emotional Robotics in the Care of Older People: A Comparison of Research Findings of PARO- and PLEO-Interventions in Care Homes from Australia, Germany and the UK
Barbara Klein, Glenda Cook, Wendy Moyle | 205

POLICY MAKING AND DISCOURSES OF AGEING

Navigating the European Landscape of Ageing and ICT: Policy, Governance, and the Role of Ethics
Eugenio Mantovani and Bruno Turnheim | 227

Ageing and Technology Decision-Making: A Framework for Assessing Uncertainty
Ciara Fitzgerald and Frédéric Adam | 257

Aging and Technology: What is the Take Home Message for Newspapers Readers?
Gregor Wolbring and Boushra Abdullah | 271

Towards an Ageless Society: Assessing a Transhumanist Programme
Martin Sand and Karin Jongsma | 291

Focusing on the Human: Interdisciplinary Reflections on Ageing and Technology
Maria Beimborn, Selma Kadi, Nina Köberer, Mara Mühleck, Mone Spindler | 311

List of Authors | 335

All that Glitters is not Silver – Technologies for the Elderly in Context. Introduction

LINDA NIERLING AND EMMA DOMÍNGUEZ-RUÉ

Our present society can be characterized by expectations of reaching old age and thus having a long life: life expectancy has constantly risen for the last 150 years in industrialised countries – mainly due to better living conditions, higher hygienic standards and improved medical care (Schwentker/Vaupel 2011; Hülsken-Giesler/Krings 2015). The high life expectancy we face today is indeed a success for our present ageing society, although it is very often negatively connoted and mainly discussed in the context of a "crisis" in demographic change. Among many similar statistical overviews, the EU Ageing Report 2012 states that "one in three Europeans will be over 65 by 2060. The ratio of working people to the 'inactive' others is shifting from 4 to 1 today to 2 to 1 by 2060" (European Commission 2012: online). The "challenge" of coping with this growing share of "inactive" population has fostered many initiatives by political, social and research institutions (see also Mantovani/Turnheim in this volume)[1]. In the last decades, the field of ageing studies has also driven the attention of scholars from many disciplines in an effort to respond to this alteration in the social structure and in an attempt to reflect and also improve the quality of life for elderly people (e.g. Kriebernegg/Maierhofer 2013; Twigg/Martin 2015). Technological advances have undoubtedly contributed to improve the lives of elderly citizens in numerous aspects and are prominently included in political ambitions to

---

1 | Especially at EU level, various initiatives focus on the topic of technology and ageing – this volume is also related to findings and actions carried out within three different EU-funded projects: SIforAge (http://www.siforage.eu), Value-ageing (http://www.valueageing.eu), and PACITA (http://www.pacitaproject.eu).

"solve the challenges of an ageing society" – ranging from IT and communication tools over specific care technologies to robot companions.

This volume is dedicated to the role of technologies in an ageing society as "technologies" are, more often than not, explicitly addressed in social and cultural discourses on ageing, aside from being underlying aspects to many facets of an ageing society, such as in the discourse on Active Ageing (Formosa 2013) or more concretely in the field of (new) media (Lövgren 2013). Furthermore we can observe that often the demographic crisis of ageing is turned "into a major 'societal challenge'" (Cuijpers/van Lente 2015: 54) which is to be addressed by technical solutions based on innovation. Empirical evidence, however, suggests that the development and use of technologies for the "silver generation" is not as smooth as it is intended to be as regards unredeemed return of investment (Bieber/Schwarz 2011), unintended side effects (Krings et al. 2014) or unexpected social relations and emotions towards technologies (Böhle/Bopp 2014, Turkle 2012, see also Fernández-Ardèvol in this volume). The complex relationship between the ageing individual and technology was already pointed out by Charness, Parks and Sabel (2001), highlighting both the prospective benefits and the potential risks of such an interaction. More recent academic works have likewise perceived the need to shift the focus from the technological to the human aspects to understand the co-construction of the social phenomenon of ageing and the inscribed role of technology in it (Felsted/Wright 2014, Peine et al. 2015, Pelizäus-Hoffmeister 2013).

In this context, it seems crucial to continue emphasizing that the decision to whether a technical solution is "good or bad" cannot be judged from the given technology as such: it rather concerns the question of *how* a technology is integrated in the social surroundings. Furthermore, it is important to broaden one's view to differentiate the range of technologies which are relevant for the context of ageing while avoiding the risk to homogenize the group of old people (Aceros et al. 2015): relevant "aging technologies" are not only the often politically mentioned "care" technologies – such as assistive technologies for the frail elderly. Rather, a range of technological solutions play a role in earlier phases of ageing like everyday technologies, household technologies, communication technologies or biotechnologies. It shows that these also encompass widely familiar technological solutions, which are in turn approached and used by the elderly in a different way from the above mentioned, with relations to their life course and biography that may include "savvy tech-operators, ambivalent users *and* non-users." (Loe 2015: 141, see also Pelizäus-Hoffmeister and Fernández-Ardèvol in this volume).

The explication about how technologies are used by the elderly is often seen as basis to go a step further into identifying needs and offering technological solutions developed and designed for the elderly, which should specifically suit their needs (e.g. Suopajärvi 2015), a strategy that reflects the evidence that in Western countries technologies, and especially ICT, are increasingly positioned as *the* solution to the problems usually associated with ageing. Although this sounds like a very promising strategy to support policies of "active ageing" at an early stage, the role of technology in elderly people's lives, together with their needs and habits towards it, are not as clear as they should be at first sight but rather require more sensitive approaches to technological development and design (see Marston/Graner-Ray and Biniok et al. in this volume).

Indeed, a recent attempt to bridge the gap between technology and its prospective human users –especially in the field of "care technologies" – was provided by an approach that integrated scientists other than engineers as well as user groups into the technical design and development process. Recent research has, however, shown that this integration is by far not "the end of the story", but rather opens new questions as regards acceptance and integration of differing (scientific) views as well as interpretational sovereignty, and thus power structures in the collaborative design process (Compagna/ Kohlbacher 2015, see also Endter and Kamphof in this volume). Especially as regards the field of care, at a political level, we can often still observe a "technology-push" strategy that involves a high prominence of aspects like efficiency or technical feasibility, which does not meet the human complexity of the field (Hülsken-Giesler/Bleses 2015, Mol et al. 2010, Oudshoorn 2011, Weinberger et al. in this volume). "Human complexity" here covers a wide range of factors like personal attitudes towards technology as well as the context in which the technology is embedded, be it the underlying conception of 'care' of the affected groups, the grade of frailty of the elderly (e.g. dementia), or the (national) surrounding in which certain types of technology are introduced (López Gómez 2015, see also Klein et al. in this volume).

Furthermore, technology-based care interventions cause a close engagement of technology with the lives of the elderly without being at all "neutral". Rather, as Peine/Moors put it, "they redefine boundaries (between health and disease, between citizens and patients, between what is considered active and what is not), they limit or enable (sometimes in unexpected ways) agency in one way or the other, and they define new socio-technical arrangements in which responsibilities, actions and interactions are re-distributed among existing and new stakeholders involved with health and care" (2015: 69).

The above mentioned technological changes as regards boundaries, agency and the socio-technical arrangement as such are relevant for both types of elderly care – stationary and ambulant care. To illustrate this with an application from ambulant care, a widely used and allegedly "simple" technological solution is alarm pendants. Despite their apparent simplicity, empirical research shows how resilient elderly people are to the introduction and use of this technology in practice (e.g. Pritchard/Brittain 2015; López Gómez 2015) and also how the device changes the close private context of the elderly, that is, the "home" in terms of "boundaries" and "agency" of the elderly. The implications of technology use in this field thus go beyond the socio-technical arrangement as such. The case of the alarm pendant can also be used to show the societal contradictions connected with technology development and use "for" an ageing society, as there are a range of underlying discourses leading to the practical introduction of an alarm pendant which are very seldom questioned; first, a discourse in "ageing-and-innovation", and second, a discourse implying that "older-people-want-to-live-at-home". As Neven (2015: 39) points out, these two discourses powerfully merge and strongly guide technology development for the elderly. First, the "ageing-and-innovation discourse" usually states that increasing costs for care, coupled with scarce and overworked personnel and too few places in care homes, pose a major societal problem that has to be solved with technologies. The second discourse Neven uncovers is what he calls an "older-people-want-to-live-at-home" rhetoric (Neven 2015: 43). Both form strongly moral discourses, as they seem to be evidently and undoubtedly "right" and commonly shared by all members of society: a technical innovation allowing older people to live at home – "the preferred 'good' place to live" – at the same time solves a major societal problem and becomes thus the "'evidently right thing to do'" (43). However, this involves the danger that while older people are able to stay at home longer on their own, several factors of social relevance become neglected. For instance, social isolation or loneliness can occur while staying at home with the technology; similarly, and due to the influx of technologies, a loss of control and privacy can arise so that – through the living practice with alarm pendants – the sense of "home" itself might get lost (see also Guihen in this volume). It is thus important to explicate all occurring changes for the elderly through technologies, as seemingly easy technical innovations might be regarded immediately as 'the right thing to do'" (Neven 2015: 43), while overlooking that technological solutions need to be context-specifically assessed and sometimes need to be altered, as they might not be the best solution for each case.

As mentioned before, dominant discourses of ageing seem to foster a specific understanding of "ageing and technical innovation" at a societal level. It is thus important to shed light on the question of how discourses develop and how they are shaped by societal visions and further distributed through the media. This may concern implicit forms of discrimination of the elderly by not taking up issues publicly, e.g. concerning special age groups. This may also occur if the foundations and roots of thoughts are not made explicit and if current social concepts of ageing are built on a discourse where underlying but guiding visions like "human enhancement" are not reflected upon, which may lead to a narrow "ageist" scope of debate (Faulkner 2015, see Wolbring/Abdullah and Sand/Jongsma in this volume).

Technology policy has already reached the social sector since the last few years, as politically used-terms like "Personal Health System" (Peine/Moors 2015) for ICT based health care services, or the term "Welfare Technology" used in Scandinavia (Östlund et al. 2015) impressively imply. Policy making at European and also at national political levels in this field requires a complex interplay of sometimes opposing strategies and institutions (see Mantovani/Turnheim and Fitzgerald/Adams in this volume). It is therefore important to define – based on societal consensus – which future should be envisioned and approached in relation to technology in an ageing society. It hereby seems crucial to widen the scope towards an approach that interrogates innovation in an ageing society and not only focusses on new kinds of technologies to be developed and designed. Rather, there should be a renewed focus on the agency of politics, societal actors and the elderly themselves. Creating normative visions about what future our society is heading to is therefore crucial in this context (see Hülsken-Giesler/Wiemann 2015, see also Beimborn et al. in this volume).

The above considerations might conclude that the answers to questions that concern the ageing population are not simply new technologies but, on the one hand, a new focus on the agency of the elderly that encompasses their use of the existing technology repertoires and also their everyday strategies to maintain autonomy. On the other hand, societal actors should keep an awareness of the overall goal of technology development and design in their agendas, which has no end in itself but rather aims at serving human beings above all: "Elders [...] remind us that ultimately they aim to achieve something akin to comfortable ageing – a lifestyle that emphasizes ease, familiarity, and prudence [...]. Technology may or may not deliver comfort or control in their lived experience." (Loe 2015: 145).

Developing on these notions, this book focuses on human interaction with technology in different settings and fields and may thus help us to become more sensitive to the ambivalences it involves. It may as well enable us to adapt technologies to the people and the lives that have created the need for its existence, thus contributing to improve the quality of life of senior citizens. Hereby the scope of this volume is purposefully a wide one: it addresses both different technological fields – such as ICT or robotics – as well as different age groups – the so called "active agers" and frail elderly people in need of care. It involves not only empirically-driven case studies, but also theoretical research papers that specifically attempt to examine this multifaceted interplay between technology and ageing users.

Our volume is divided into three main parts. The first part "Ageing, Technology and (Inter-)Personal Development: Old agers as Technology Users" presents case studies that examine the relationship between various technological developments existing in everyday life (internet, mobile phones, video games or social networks) and ageing users, together with the conditions and meanings ascribed to both these technologies and its users. The second section "Ageing, Technology and Elderly Care: Assistive Technologies", explores the often complex relationship between care technologies and its users, that is, ageing patients, professional caretakers and caring relatives. The last section "Policy making and Discourses of Ageing" interrogates the role of policy makers and observes public and scientific discourses informing the relationship between technology and the elderly.

Despite the undeniable reality that our lives are surrounded and supported by an ever-increasing amount of technological developments and that the ageing population represents an important niche in the technological market, many complexities arise when observing the development of everyday technologies addressed to the elderly and their end users, which are – at least partly – addressed in the first part of the volume, "Ageing, Technology and (Inter-)Personal Development: Old agers as Technology Users". Even when the use of certain technologies is pervasive – as is the case of computers and mobile telephones –, and even when ageing/aged users represent an important target group for the commercialization of such technologies, user response might be ambiguous, contradictory, or even confrontational. The reasons for this response are multifaceted and may involve a wide variety of factors. As an example of this, *Helga Pelizäus-Hoffmeister's* qualitative study reveals the various motives for and against the use of different technological devices by the elderly to make hypotheses about the extent to which the individual, social and cultural contexts of the users influence their interpretation of technology. The author concludes that bringing in a broad understanding of interpretation and meaning into technology development and innovation is necessary, since at present the focus is much more on technical viability while the view of the user tends to be disregarded. She therefore argues that quantitative approaches which could systemically inform future technology development are still much needed.

A further qualitative evidence of this is provided by *Mireia Fernández-Ardèvol*, who studies the "non-use" of mobile phone among senior citizens (+60) in five cities around the globe with the aim of identifying common patterns of mobile non-use in different social, cultural and economic contexts. She states that "heterogeneity" best characterizes the use and also non-use of mobile phones at this age group in all the cities studied: an acceptance or rejection of mobile phone use is not a static decision. Furthermore, decisions for/against the technology in this context are actively taken, being forms of personal agency in times where one's own decisions are more and more difficult to state.

While mobile telephony is widely used by a high percentage of the population in all ages, a less visible reality is that of female elderly citizens as users of a typically youth- and male-oriented technology such as video games. *Hannah Marston* and *Sheri Rainer-Ray* focus on gaming technology and intergenerational game play: while in the 21st century gaming witnessed a change and broadened its scope to older adults, there is a gap in awareness surrounding game preferences of older women (aged 50+ years). In order to address this gap, a series of workshops were conducted to identify the type of game content older adults would like to play: these are broadly based upon their hobbies and interests, while they interestingly uncover female interests for virtual role plays that allow them to experience emotions and roles different from their everyday lives.

As the previous examples have illustrated, understanding the heterogeneity of contexts and situations in which the elderly live is crucial to enable their access to the resources and services made available to them and, by extension, their participation in society. As the availability of health care, services and entertainment opportunities is largely concentrated in urban settings, access to these in more scarcely populated areas might be limited or inexistent, thus increasing the risk of isolation and social exclusion. *Peter Biniok, Iris Menke* and *Stefan Selke* attempt to bridge this gap by exploring the use of technological devices and platforms for social inclusion of elderly citizens living in rural areas. The authors present a case study undertaken in the German Black Forest region that develops communication tools in the frame of a user-oriented online platform, which allows further social inclusion and thus a better quality of life for the elderly in that area. With the aim of making the life of senior citizens in rural areas more autonomous, with special emphasis on social inclusion, the authors meet the challenge of putting human beings first instead of technology, starting with a needs assessment that is also aimed at learning about unexpected needs and demands of the elderly (60+) with respect to web-based networking tools.

Our second section, "Ageing, Technology and Elderly Care: Assistive Technologies" is dedicated to assessing the relevance of the variety of potential settings, contexts and situations involved in technologies addressed to the elderly population. Such aspects might become visible when it comes to ensuring comfort and safety while maintaining independence as old agers face increasing health and mobility problems. It thus explores the often complex relationship between such technologies and its users - that is, ageing patients, professional caretakers and caring relatives.

At a national as well as at an international level, several innovation strategies focus on the integration of technology for a higher "efficiency" in care. In its Digital Agenda for Europe, the European Commission's Policies for "Ageing Well with ICT" show awareness of the potential of ICT in assistive technologies for elderly care, as they prognose that government spendings on pensions, healthcare, long-term care, unemployment benefits and education in the EU "will increase by almost 20 per cent, while expenditures for long-term care will double" (European Commission 2012: online). Technology-based research and innovation projects aimed at coping with this expected increase included – inter alia – fall prevention, integrated care, and the AAL Joint Programme. *Cordula Endter* provides an illustration of the materialisation of such policies in the German context by presenting an ethnographic case study that focuses on the development process of smart assistive technologies, also called Ambient Assisted Living (AAL). By conducting fieldwork in different settings – laboratories that design AAL, companies that provide this service, and households equipped with this technology – the author seeks to understand the social and cultural impact of AAL on the aged population Through an understanding of the interaction of engineers and actors in the design process as a practice, the author reconstructs the transformation of images of the elderly and their needs into a yet invisible technology. Once such technologies have been developed and brought into the market, they are implemented in different contexts, together with their interaction with the needs of the users they are intended to address. Similarly, *Barry Guihen* provides conceptual evidence of the intricacies of that process by observing the use of ICT in developing smart homes for the elderly to support the often strongly longed for ageing in one's own home ("ageing in place"), while revealing the main social and psychological implications involved in the use of such technologies. The author focusses on the human factors and emotions and cautiously elaborates on subjective aspects such as, for example, social isolation, loneliness and frustration, which are often overlooked by those involved in developing ICT-based solutions for supporting ageing in place.

So far we have examined examples of elderly citizens as willing and voluntary users (or not) of technology. In the case of the old agers (80+) and those with an incapacitating illness, issues such as vulnerability, health and safety enter in direct conflict with ethical concerns about privacy and individual consent. The ethnographic study by *Ike Kamphof* provides insights into this controversial matter by observing the role of activity monitoring as a resource used in homecare with people suffering from dementia. She analyses the frail balance between safety and invasion of the patients' privacy when technologically-mediated monitoring is used to enhance their quality of life and prevent potential hazards. Further exploration into this delicate topic is provided by *Nora Weinberger, Bettina-Johanna Krings* and *Michael Decker*, whose qualitative work incisively evidences the challenge that the increasing number of patients with dementia presents in elderly homes. The authors argue that technological developments designed for the elderly often ignore those directly involved in its use. In the case of dementia, this means that the mobility of people should be enabled, while independence must be limited to avoid potential danger for the patient. By proposing, applying and reflecting on an approach that focuses on the needs of (dement) users, the authors strengthen the necessity to re-embed dement patients into social spaces that can cope with their special condition.

All of the authors presented so far agree in the fact that those directly involved in the use of technology in cases of patients with dementia should participate in the implementation and – ideally – in the design of technologies that better address the patients' needs. Likewise, they emphasize that their feelings towards and perception of such technologies need to be taken into account. Evidence on how a technology in a care setting can be evaluated is provided by the joint work by *Barbara Klein, Glenda Cook* and *Wendy Moyle*. The authors present findings from a study of robot therapy with the already classical baby seal "PARO" in three different national research settings (Australia, England and Germany) and demonstrate that interaction with emotional robots can have a positive impact on patients with cognitive impairment when they are provided together with other activities in elderly homes. The authors, however, stress the necessity to continue research on emotional robots as many questions in this field are still not answered.

Given the evidence provided so far, the complexity and variety of interactions between the elderly and the technologies addressed to them reveals an impending need to implement policies in relation to these issues. Our last section, entitled "Discourses of Ageing and Policy Making", thus focuses on current social debates on ageing and how they are applied – or ignored – by governments and institutions when implementing policies addressed to the senior population.

A useful way of navigating among the sea of troubles (and documents) of EU regulations is provided by *Eugenio Mantovani* and *Bruno Turnheim*, who give a comprehensive view on existing European policies on ageing, together with the extent to which public policies are representative of and informed by the views of those agents who represent elderly citizens or by elderly population at large. The authors historically analyse the development of policies for ICT and active and healthy ageing within the EU framework and its related agencies – regulatory actors and agencies, interest groups, research centres, etc. – with an emphasis on whether ethical and moral considerations are taken into account (or not). The authors conclude that current political approaches still lack public deliberation processes, a fact that would be detrimental to reach social legitimacy in this field.

As a complement to this data collection, Ciara Fitzgerald and Fred Adam systematically examine the uncertainties that decision makers face when implementing European telecare policies in the context of a growing ageing population. Based on an analysis of national approaches to telecare in 14 European countries, the authors differentiate uncertainties in the field of policy making for telecare along the categories of technological mechanics, impact, and societal preferences. The authors argue that such transparent decision support tools offer helpful guidance to decision makers, especially as they will increasingly be challenged to consider a framework of 'responsible innovation' in their policy making efforts.

Institutional discourse on ageing and technology inevitably informs institutionalised attitudes towards it: in this sense, the media provide an apt mirror that reveals perceived notions on the matter. The work by Gregor Wolbring and Boushra Abdullah quantitatively analyses the coverage of the topics of ageing and technology in two Canadian newspapers. By revealing when, where, and in which way technologies are mentioned in relation to ageing, their study seeks to reveal existing media discourses about ageing and technology and discuss it jointly with Canadian key policy documents on aging. Furthermore, in a critical intent, the authors also detect which issues are not taken up by the media, namely the socially disadvantaged elderly such as elderly with disabilities, elderly immigrants and the poor elderly. They argue that the media should take up its educational mission seriously and not only cover these neglected issues, but they should also include relevant policy documents in their reporting. Scientific-philosophical trends on the phenomenon of ageing likewise offer a comparative view that provides an insight into the discourse of ageing in relation to the advances of technology. Martin Sand and Karin Jongsma discuss the vision of Transhumanism, a highly controversial trend of thought that aims at overcoming ageing, illness and mortality in the face of unprecedented technological development. With both philosophical and technological arguments, the authors interrogate the ethical validity of the notions proposed by the transhumanist approach and their goals of research on anti-ageing. Instead they propose a wider understanding of the phenomenon of ageing, not as a simply biological course of physical and mental degradation but as a complex social process.

In view of the many facets involved in the relationship between technological developments and the elderly, it is necessary to take into account a number of ethical issues when it comes to developing and implementing technologies addressed to the elderly. We thus come full circle and return to our initial question of what it conceptually means to take the "human factor" first. A theoretically-driven ethical reflection is provided by Maria Beimborn, Selma Kadi, Nina Köberer, Mara Mühleck and Mone Spindler, who examine the notion of "focusing on the human" from a sociological view point, followed by a clear statement concerning the normative implications of "good life" and "good care" with technologies. The authors conclude by stating the need for thorough perspectives on the topic, which would not only integrate different scientific fields but also affected citizens in a true transdisciplinary approach.

The articles in this volume have illuminated the complexity of the human factor in the attempt to technically approach the "challenges" of an ageing society. They shed light into the various political, social and personal arenas where the encounters between technologies and elderly people take place. All the contributions set the "human factor" at the center of their analysis and describe the specific outcome of this relationship through sensitive approaches, be it observed in subjective processes of identity formation - how respect and dignity for the elderly can be found in public discourse - or in the transformation of age-related issues into policy processes. Given that all contributions focus on the human factor through various facets, it is high time to abandon the idea that the ageing individual is still seen as a malfunctioning machine whose deficiencies must be diagnosed or as a set of limitations to be overcome by means of technological devices.

To this purpose, the above mentioned contributions open further desiderata for research in the field of technology and ageing. It thus seems important to continue focusing the research on clear differentiations in this field, distinguishing different age groups and different technologies to enable assessments and view on the different roles technologies can take in a specific context. What seems however crucial for the future is to bridge the gap still present between sensitive evaluations and assessments on a context-specific individual level and political solutions guiding further technical innovations, organizational practices and institutional settings in this field.

This book could not have been published without the support of many contributors. Our biggest thank you goes to the authors, whose excellent work has provided life and flesh to our initial idea and without whose effort this book would not have been possible. We would also like to thank our respective institutions, the University of Lleida (Spain) and the Karlsruhe Institute of Technology (Germany) for providing us with the support – financial and logistic – to materialize the volume. The EU project SIforAge was also very kind in partially funding this project. We cannot but mention our publisher, transcript Verlag, and very especially Annika Linnemann, whose advice and constant assistance for many queries eased the publishing process. Last, but not at all least, we would like to express our gratitude to Gabriele Petermann and Marei Lehmann for their reliable layout and formatting work and proof of bibliography. Their help has been absolutely instrumental for us. Finally, we would like to dedicate this book to our respective first daughter and son who were both born and have developed together with this book – of less assistance for the finalizing process, but making their own contribution to demographic change from the other side of the pyramid.

REFERENCES

Aceros, J.C./Pols, J./Domènech, M. (2015): "Where is grandma? Home telecare, good ageing and the domestication of later life" In: Technological Forecasting & Social Change 93, pp. 102-111.

Bieber, D./Schwarz, K. (eds) (2011): Mit AAL-Dienstleistungen altern. Nutzerbedarfsanalysen im Kontext des Ambient Assisted Living. Saarbrücken: iso-Institut.

Böhle, K.; Bopp, K. (2014): „What a vision: The artificial companion. A piece of vision assessment including an expert survey". In: Science, Technology & Innovation Studies (STI Studies) 10/1, pp. 155-186

Charness, N./Parks, D./Sabel, B. (2001): Communication, Technology and Aging: Opportunities and Challenges for the Future. Berlin: Springer.

Compagna, D./Kohlbacher, F. (2015):The limits of participatory technology development: The case of service robots in care facilities for older people. In: Technological Forecasting & Social Change 93, pp.19-31

Cuijpers, Y./van Lente, H. (2015): "Early diagnostics and Alzheimer's disease: Beyond 'cure' and 'care'." In: Technological Forecasting & Social Change 93, pp. 54-67.

European Commission (2012): Digital Agenda for Europe - the European Commission's Policies for Ageing Well with ICT: Retrieved from http://ec.europa.eu/digital-agenda/en/policies-ageing-well-ict (last accessed 15.09.2015)

European Commission (2012): EU Ageing Report 2012: Retrieved from https://ec.europa.eu/digital-agenda/en/news/2012-ageing-report-economic-and-budgetary-projections-27-eu-member-states-2010-2060 (last accessed 15.09.2015)

Faulkner, A. (2015): "Usership of regenerative therapies: Age, ageing and anti-ageing in the global science and technology of knee cartilage repair." In: Technological Forecasting & Social Change 93, pp. 44-53.

Felsted, K.F./Wright, S.D. (2014): Towards Post Ageing: Technology in an Ageing Society. Berlin: Springer.

Formosa, M. (2013): "Positive Ageing in an Age of Neo-liberalism. Old Wine in New Bottles?", In: Kriebernegg, U./Maierhofer, R. (eds.), The Ages of Life. Living and Ageing in Conflict? Bielefeld: transcript, pp. 21-35.

Hülsken-Giesler, M./Bleses, H.M. (2015): "Schwerpunkt: Neue Technologien in der Pflege". In: Pflege & Gesellschaft 20 (1), pp. 3-4.

Hülsken-Giesler, M./Wiemann, B. (2015): "Die Zukunft der Pflege – 2053: Ergebnisse eines Szenarioworkshops". In: Technikfolgenabschätzung Theorie und Praxis 24/2, pp. 46-57.

Hülsken-Giesler, M./Krings, B.-J. (2015): "Technik und Pflege in einer Gesellschaft des langen Lebens. Einführung in den Schwerpunkt". In: Technikfolgenabschätzung Theorie und Praxis 24/2, pp. 4-11.

Kriebernegg, U./Maierhofer, R. (eds.) (2013): The Ages of Life. Living and Ageing in Conflict? Bielefeld: transcript.

Krings, B.-J./Böhle, K./Decker, M./Nierling, L./Schneider, C. (2014): "Serviceroboter in Pflegerarrangements." In: Decker, M./Fleischer, T./ Schippl, J./ Weinberger, N. (eds.), Zukünftige Themen der Innovations- und Technikanalyse: Lessons learned und ausgewählte Ergebnisse. Karlsruhe: KIT Scientific Publishing, pp. 63-121.

Loe, M. (2015): "Comfort and medical ambivalence in old age". In: Technological Forecasting & Social Change 93, 141-146

López Gómez, D. (2015): "Little arrangements that matter. Rethinking autonomy-enabling innovations for later life" In: Technological Forecasting & Social Change 93, pp. 91-101.

Lövgren, K. (2013): "Celebrating or Denying Age? On Cultural Studies as an Analytical Approach in Gerontology." In: Kriebernegg, U./Maierhofer, R. (eds.), The Ages of Life. Living and Ageing in Conflict? Bielefeld: transcript, pp 37-55.

Mol, A./Moser, I./Pols, J. (eds.) (2010): Care in Practice. On Tinkering in Clinics, Homes and Farms. Bielefeld: transcript.

Neven, L. (2015): "By any means? Questioning the link between gerontechnological innovation and older people's wish to live at home." In: Technological Forecasting & Social Change 93, pp. 32-43.

Östlund, B./Olander, E./Jonsson, O./Frennert, S. (2015): "STS-inspired design to meet the challenges of modern aging. Welfare technology as a tool to promote user driven innovations or another way to keep older users hostage?" In: Technological Forecasting & Social Change 93, pp. 82-90.

Oudshoorn, N. (2011): Telecare Technologies and the Transformation of Healthcare. Palgrave macmillan: New York.

Peine, A./Faulkner, A./Jaeger, B./Moors, E. (2015): "Science, technology and the 'grand challenge' of ageing – Understanding the social-material constitution of later life." In: Technological Forecasting & Social Change 93, pp. 1-9.

Peine, A./Moors, E. (2015): "Valuing health technology – habilitating and prosthetic strategies in personal health systems." In: Technological Forecasting & Social Change 93, pp. 68-81.

Pelizäus-Hoffmeister, H. (2013): Zur Bedeutung von Technik im Alltag Älterer: Theorie und Empirie aus soziologischer Perspektive. Berlin: Springer.

Pritchard, G.W./Brittain, K. (2015): "Alarm pendants and the technological shaping of older people's care. Between (intentional) help and (irrational) nuisance." In: Technological Forecasting & Social Change 93, pp. 124-132.

Schwentker, B./Vaupel, J.W. (2011): "Eine neue Kultur des Wandels." In: Aus Politik und Zeitgeschichte 10-11/2011, pp. 3-10.

Suopajärvi, T. (2015): "Past experiences, current practices and future design. Ethnographic study of aging adults' everday ICT practices – And how it could benefit public ubiquitous computing design." In: Technological Forecasting & Social Change 93, pp. 112-123.

Turkle, S. (2012): Verloren unter 100 Freunden. Wie wir in der digitalen Welt seelisch verkümmern. München: Riemann.

Twigg, J./Martin, W. (eds.) (2015): The Routledge Handbook of Cultural Gerontology. Routledge: London

Ageing, Technology and (Inter-)Personal Development: Old Agers as Technology Users

Motives of the Elderly for the Use of Technology in their Daily Lives

HELGA PELIZÄUS-HOFFMEISTER

The subject of aging and technology has become increasingly important in both the research and public sectors. Many experts are convinced that some of the challenges caused by the demographic shift can be overcome with the use of technical equipment. However, technological development is often orientated more on innovation than on user demands. The focus is on the feasibility of the technology, while the perspective of the user is often a neglected factor. One of the consequences of this neglect is that many products which are created for the elderly are in turn rejected by them (e.g. Technische Universität Berlin 2011; Pelizäus-Hoffmeister 2013). Based on the findings and concepts of several sociological technology studies (see Rammert 2007, 1988; Wengenroth 2001; Woolgar 1991; Hörning 1988, 1989), one may assume that a device will only be accepted by the elderly and integrated into their daily lives when their interpretations of technology are taken into account during the technology's development (cf. Giesecke 2003: 10).

This chapter focuses on the meanings and ideas which elderly often connect with the devices in their daily life. The findings presented here are a result of a qualitative-orientated research study conducted in Munich (cf. Pelizäus-Hoffmeister 2013). The aim of this study was to describe and explain the different interpretations of technology by older people, while taking into account their individual behavior patterns as well as the social, structural, cultural, and individual context conditions. The purpose of conducting this analysis was to develop hypotheses regarding relations between context conditions and the interpretations of technology by the elderly.

The chapter is divided as follows: First, the four theses of Hörning (1988, 1989) about technology and its cultural meanings (1) will be explained. They are used as sensitizing concepts in the empirical aspect of this study. Next, the study itself will be described, and the methodological approach taken will be explained (2). After that, the results in the form of a typology (3) will be presented. In the conclusion, the results will be evaluated in the context of future technological development (4).

1. CULTURAL ASPECTS OF TECHNOLOGY

Since the 1980s, the topic of technology in everyday life has been paid increasing amounts of attention from researchers in the field of sociology. Two different research perspectives describe the relationship between technology and people in everyday life. The first perspective is focused on the structuring and regulating effect of technology in daily life, as described by Habermas' (1981) concept "colonization of the 'Lebenswelt'" (e.g. Bievert/Monse 1988; Joerges 1988). From the other perspective, technology is described as an element of the cultural sphere (e.g. Hennen 1992; Hörning 1988, 1989; Rammert 1988, 2007). Here it is emphasized that the use of technology is based less on its functions than on the meanings associated with it (cf. Hörning 1988). These perceptions are developed on the one hand by the people who produce, evaluate, and disseminate technology. On the other hand, they are also formed by the people who use the technology in their everyday lives (Hörning 1989: 91). In consequence, technology has lost its character as solely an instrument. Rather, consumers of technology associate with it a variety of individual meanings which do not necessarily match those of the producers. Their meanings are partially affected by culture and society, and also by their everyday applications (ibid: 117). From Hörning's perspective, devices cannot be attributed with meanings merely due to their materiality or functionality. He emphasizes that technology is interpreted with other meanings in addition to their strictly functionalist meanings. These meanings, produced by the users themselves, decide whether the device is purchased and used or not.

If this research perspective is recognized as a requirement in developing user-friendly and accepted technology for the elderly, we have to ask: How do various groups of elderly persons interpret technology based on their use in their everyday lives? Or more specifically: What are their motives for and against the use of technology?

Hörning describes four general models of orientation. In his opinion, they comprise the basis of using devices (cf. 1988: 73). First, he outlines the so-called *Control Orientation*. In this sense, technology is used to control the environment, or at least allow the user to have the feeling that it can be controlled if necessary. The desire for joy and pleasure in the use of technology – for example motorcycling, to feel the "great freedom" – is labelled as an *Aesthetic-Expressive Orientation* (cf. ibid: 76). *Cognitive Orientation* is characterized by the desire to use technology as a means of empowering the user in dealing with the technical environment. According to Hörning, this interest is rooted in social pressure, using technology to appear "rational" or "intelligent" (ibid: 77). Lastly, Hörning attaches a special importance to the *Communicative Orientation*. He argues that in this orientation, the use of technology enables the user to integrate into a variety of communication systems and allows them to avoid marginalization in their social environment (ibid: 78). In summary, it can be assumed that the four models of orientation are often associated with each other, although it is possible that they differ in the individual depending on the specific device.

In this study, Hörning's theoretical assumptions are used as sensitizing concepts, as described by Kelle and Kluge (1999: 25). They help to draw attention to the different and varied meanings of technology to the elderly, but they are not interpreted as limitations to the aspects described above.

2. METHODICAL APPROACH

The author of this study interviewed thirty-one men and women. All respondents were at least sixty years of age (cf. Pelizäus-Hoffmeister 2013). Although it is clear that biological age is an insufficient metric, this age limit was chosen for pragmatic reasons based on the definition of the World Health Organization (see WHO 2002: 4). There are significant differences between people of the same age in their health, participation, degree of independence, etc. (cf. ibid.). It is also worth noting that the number of women interviewed was slightly greater than that of men.

The interview method developed by Kaufmann (1999), the "verstehende" interview, was used to conduct this study, in addition to a written questionnaire (cf. Hopf 1991: 177; Witzel 2000). Based on the narrations of those interviewed, the different forms of technology use are described here, as well as the meanings which interviewees associated with technology. From this, the aim was to discover the individual, social, structural, and cultural conditions that lead to a successful or unsuccessful utilization of technology. The questions were limited to the use of technology in the home of the elderly, on the one hand to keep this complex issue workable, and on the other hand due to the high value of the home for the elderly (see Kreibich 2004: 12; cf. Backes/Clemens 2003: 230; Saup 1993).

The interviews were digitally recorded and then transcribed. The data analysis was based on the Grounded Theory Method, developed by Strauss and Corbin (1996). The aim in doing so was to construct a *Gegenstandsorientierte Theory*, which identifies interrelationships between meanings of technology, the use of technology, and other context conditions. The author of this study developed a typology from a comparative analysis of the individual cases. Although the results are not statistically representative, they could potentially show how the user-associated meanings of technology influence the use of technology and vice versa. The types are not to be understood as personal types, but rather as interpretative models (*Deutungsmuster*). Accordingly, a person can be assigned to several types if he or she follows different motives in the use of technology.

3. INTERPRETATIVE MODELS OF TECHNOLOGY – MOTIVES FOR AND AGAINST THE USE OF TECHNOLOGY

In this study, Hörning's theoretical assumptions were used as sensitizing concepts. The interpretative models of technology being developed – respective to the motives of the elderly for or against the use of technology – were assigned to the following more general dimensions, which partly correspond with Hörning's orientation models: the Instrumental Dimension, the Aesthetic-Expressive Dimension, the Cognitive Dimension, and the Social Dimension. This may be seen in Figure 1 below.

The interpretative models of technology were constructed as action types in order to more effectively outline their specific characteristics. They are not to be understood as types of people, but rather as typical actions and interpretation models. Eight different models to represent the motives of elderly for and against the use of technology were developed.

The Instrumental Dimension includes the models *Technology as an Invisible Hand, Saving with the Selective Application of Technology,* and *Technology as a Blessing*. These three models of elderly classified in this dimension are focused especially on the labor-saving functions of technology. Here, technology is seen as an instrument that can improve or support the user's natural abilities and skills. Through the use of technology, housework can be done faster, easier, more efficiently, and more economically. The model *Pleasure in Beautiful Technology* belongs to the Aesthetic-Expressive Dimension. For this user type, the use of technology is often associated with the emotions of joy and pleasure. *Control over Technology* and *Technology as a Passion* are models which belong to the Cognitive Dimension. These are quite different from the models which fall in the first two general dimensions. The occupation with technology itself rises to a prominent role for these users. The aim of the elderly is, for example, to understand the options and functions of technology, and in this way train their cognitive skills. The Social Dimension also includes two models from this study. One is named *Construction of the Social Self with Technology*, and the other is titled *Communication with the Aid of Technology*. The focus of the users here is on his or her own social needs. On one hand, the use of technology can assist the elderly to feel integrated into their social circle, who are also generally interested in new technology. On the other hand, the user can be connected with their greater social network using new communication media.

Figure 1: Interpretive Models of Technology

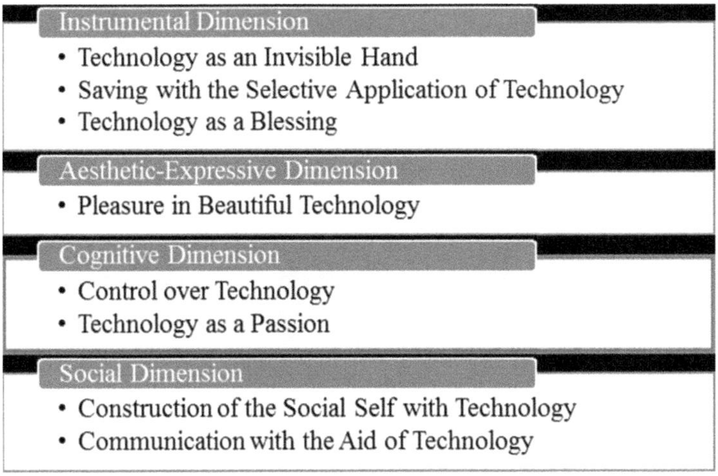

Instrumental Dimension
- Technology as an Invisible Hand
- Saving with the Selective Application of Technology
- Technology as a Blessing

Aesthetic-Expressive Dimension
- Pleasure in Beautiful Technology

Cognitive Dimension
- Control over Technology
- Technology as a Passion

Social Dimension
- Construction of the Social Self with Technology
- Communication with the Aid of Technology

These interpretive models are presented as follows: First they are described briefly. Then the – not always conscious – motives and actions based on these categories are presented. Lastly, the author discusses the search for links between these actions, interpretative models, and the context conditions such as generation, gender, education, social participation, etc.

Technology as an Invisible Hand

The respondents, who were assigned to this model, see the use of technology as a chance to do their housework better, faster, and easier. They also assess technology as merely a means to an end. In their eyes, it has no value in itself. Technology should be like an invisible assistant, expressed in a metaphorical sense. It should not be the focal point in their everyday lives. The normative character of this interpretative model points out the desire of the elderly to minimize their handling of technology.

What are the motives behind this attitude? Two main motives (I, II), which are congruent in a certain sense, but also differ to some extent, are evident for those who use technology as an "Invisible Hand." The people of both subtypes have in common that they wish to devote as little time as possible to conducting their housework. They prefer to spend their time with other activities such as meeting friends, reading, walking, or playing with their pets. If technology can be helpful to do the "annoying everyday stuff" more quickly, as some participants put it, it is accepted and incorporated into daily life. These elderly associate technology with their unpopular housework. Therefore, it stands in contrast to the "brighter side of life," and should require as little time and space as possible.

For the people of Subtype I, another reason exists for the wish of "invisible" technology. Especially with the use of new technology, these elderly people often associate many difficulties, troubles, and a lack of their own competence in dealing with it. The elderly of Subtype II, however, are convinced of their own high expertise in dealing with technology. At the same time, they are not interested in technology in itself, so they avoid it as well, if possible, despite their perceived proficiency.

Based on this interpretative model, the elderly with these motivations use particularly well-known and simple[1] devices in their daily lives. The use of new devices is avoided if possible, because they are associated with a certain degree of involvement with the technology itself. Moreover, the elderly of Subtype I avoid new devices because they do not want to be confronted with

---

1 | Remarkable is the difference in their identification of "simple" devices. People who are convinced that they have little skill in using technology describe only classical technology – such as the washing machine, cooker, and mixer (cf. Tully 2003) – as simple. On the contrary, people with a lot of experiences in using technology add the PC, the mobile phone, and the Internet to this list.

their perceived lack of skill with technology. These elderly prefer devices they have already been used in their household for a long time. As viewed through theoretical concepts, it means that the acceptance of technology is accordingly limited to the so-called *"veralltäglichte"* technology (Hörning 1988: 51). This means that these devices are often not even consciously perceived as technology because they have been effortlessly integrated in the household.

Sackmann and Weymann (1994) suppose that the interpretative models of technology differ due to the varying experiences the various age groups make in their youth. Based on this study, the model *Technology as an Invisible Hand* can be found across all age groups. Accordingly, it can be assumed that age cohorts do not play a relevant role for persons belonging to this type. In addition, the personal perception of competence in using technology seems to have no influence, because this type is represented by people of both categories of self-described competence in technology use, low and high. Despite this, we can also see that the feeling of having a lack of competence can increase within this model. Based on this, the author assumes that women particularly can be assigned to this type, because they often associate with a lack of technical competence, which in turn influences their self-image (Ahrens 2009; Beisenherz 1989; Collmer 1997; Hargittai/Shafer 2006).

Saving with the Selective Application of Technology

The functionality of technology is at the central motivation for this model as well. The use of technology must save resources of all kinds. Household work is considered from the perspective of their rationalization. The representatives of this type are constantly looking for means of optimizing this; they use technology when it appears economically. Otherwise, and preferably, they avoid it.

What motives determine this interpretative model? Here we discover a strong orientation towards values such as efficiency, performance, and thrift. These elderly persons attempt to organize their daily activities in such a manner so as to complete them as simply, quickly, cheaply, and efficiently as possible. The motive of "saving" seems almost to be an end in itself, which determines many behaviors throughout the lives of those associated with this model (see also: Rammert 1988: 192). For them, the use of technology makes sense only if it helps to save time, money, energy, effort, or other resources. However, their main focus seems to be using technology only when its use cannot be avoided. How can this be explained?

The motive of rationalization implies a strong desire for control. From the perspective of these elderly, their ability to control things is threatened by the use of technology. They perceive technology as a threat to their control options, as it would determine their daily schedules, and therefore their whole lives. Especially with new, digital, and complex technologies, these older individuals associate a variety of risks which limit their control. For example, one interviewee claims to avoid the use of computers because he is afraid of becoming overwhelmed by its many functions. Moreover, he fears he would require a great deal of outside support due to his lack of competence in using this technology, which would further reduce his perceived control.

Nevertheless, the use of select technology in daily life remains a necessity for these elderly. They use common classical devices such as televisions, washing machines, cookers, and telephones. Even regarding these devices though, they raise occasional concerns. Especially when a device is defect and requires repair, they devote a good deal of time in thinking about ways of avoiding its replacement. These people remain skeptical in their use of technology.

Who are the people belonging to this type? These elderly have rather low technical competence, which can be traced back to a lack of professional experience with new technologies earlier in their lives. It could also be an effect of a life beyond employment, which is characterized by a low use of new technology in the majority of cases. The latter usually affects older women, who were mainly responsible for child care and housework during their working years. Accordingly, this technology skepticism and the rejection of use of new technology can be attributed to a lack of biographical experiences with it.

Technology as a Blessing

In this type as well, the functions of technology are at the center of motivation. Technology should support the individual in completing daily tasks, especially those requiring high amounts of physical stress. In addition, the elderly associated with this model assess technology as absolutely positive. Moreover, it is worth noting that technical problems do not play a pivotal role in their daily lives. These elderly people are convinced of their ability to master their technical equipment to a competent level.

What is the subjective basis for their absolutely positive perception of technology and their feeling of technical competence? Furthermore, what context conditions allow these interpretations? These elderly persons associate only advantages with the use of technology because they compare their lives in the past with those in the present. The past generally relates to their youth, when housework was associated with high physical stress because labor-saving devices were hardly available. Based on these past experiences, it is plausible that current household appliances with widespread, labor-saving capabilities are welcomed with great openness, enthusiasm, and appreciation.

In addition, the positive interpretation of technology can further be attributed to the fact that these older people do not desire to understand their devices' working methods or their implicit logics and functions; they only wish to utilize them. A preoccupation with technology in itself is not a factor for these individuals, with the result that many potential problems and risks which may be associated with new digital technology are not perceived. Moreover, they associate technical competence with the ability to use technology, which results in the effect that they perceive themselves as technically competent despite their actual low level of skill.

What context conditions allow this interpretative model of technology? The elderly of this type are characterized by the fact that they have a reliable social network, which provides them with necessary devices. Moreover, it ensures the elderly can use the devices without significant issues. When problems do occur, their social network ensures fast troubleshooting.

On this basis, it is not surprising that these elderly use a variety of both classic and new digital devices. One older female interviewee, for example, often carries a mobile phone when she walks her dog in order to call for help if necessary. She also uses the clothes dryer, an example of classic technology, to avoid having to hang dry her laundry. All this time, she utilizes the devices without apprehension because she is convinced that she is able to use them competently.

In this study sample, this interpretative model was represented only by seniors of advanced age. Sackmann and Weymann (1994) call them the "pre-technological generation." These people can be characterized by the fact that, in their childhood, electricity and electric light was already present in households, but radios were the only complex device they knew for many years. In this respect, it is obvious that the above-described comparison between the past and present leads to their conclusion that labor-saving technology is always an enrichment in their lives.

The very old age of the respondents in this model may also be a result of the fact that they especially appreciate devices which reduce physical effort. This is likely because the occurrence of physical impairments increases with age. All respondents of this type possessed little technical knowledge, which is also due to their advanced age and therefore fewer opportunities to learn how to use the new digital technologies. It could as well be the reason for their lack of desire to understand the inner logic of new technology.

Pleasure in Beautiful Technology

This model, which is attributed to the Aesthetic-Expressive Dimension, is rarely represented among elderly individuals. For those that do associate with this type, the functions of devices are of great importance, as in the Instrumental Dimension, but they nevertheless attach a great importance to the appearance of the device as well. Each device is expected to contribute to the beauty of the individual's home. Here, beauty is understood in the sense of a positive feeling or event.[2] Therefore, not only the functionality but also the shape, color, and design are important when buying a coffee maker, for example.

What motive can explain this interpretative model of technology? For these elderly, aesthetics are generally of great importance, which is also reflected in their style of dress and in their home environment. The visually pleasing arrangement of all the artifacts of their home initiates their experience of pleasure, joy, and well-being. At the same time, they are convinced that the choice of a – usually expensive – device underlines their wealth and their cultural expertise, which in turn increases their social prestige.

Based on this motive, the elderly in this model spend a lot of time and give a great deal of attention to purchasing a new device. The desire for a beautiful device can even be so strong that they dispense with functionality in order to get the best style. They appreciate high-priced, brand name products, and often prefer to make their purchases in specialized shops. The atmosphere of these shops is also in line with their desire for beauty in their way of life. In addition, they prefer new and novel products, because they associate a feeling of being up-to-date with them.

Which people could be assigned to this type? Or to ask more generally: What conditions correspond with this interpretative model of technology? It is noticeable that the respondents of this type are similar to the group of people Schulze (1992) describes in his *Erlebnisgesellschaft*, namely the

---

2 | Beauty is not an objective phenomenon, but it is associated with a device by the elderly.

Niveaumilieu. These men and women mainly have higher levels of education and are rather wealthy. One may suspect that the related cultural and economic resources are the necessary basis for this interpretative model.

Control over Technology

Characteristic for the elderly in this model is their desire to improve their technical abilities and skills via their use of technology. They often take for granted that the devices also have to fulfill important functions in the household (Instrumental Dimension). The aim of these older people is to master their technical devices. New digital technology tends to especially motivate them, as it appears to be complex and inscrutable at first glance, which makes it interesting in their eyes. Similarly, technical problems are welcomed as interesting challenges rather than as nuisances. Nevertheless, these people are not interested in all technical innovations, but rather deal exclusively with the devices of their household.

What are the motives of these people? First, the use of technology is a welcome opportunity to turn away from what these people consider to be boring daily tasks. In addition, they enjoy the opportunity to improve their cognitive abilities and their technical skills, while at the same time assuring themselves of their expertise. Along with technical competence, they associate high cultural qualifications, intelligence, and mental flexibility, all of which play an important role in their lives. Therefore, their own technical competence is a focal aspect of their self-image. One the one hand, it triggers a feeling of pride and a sense of being pleased with themselves. On the other hand, it allows them to set themselves apart from people who possess little technical expertise. For these people, technical competence is the basis of social affiliation, and at the same time an indicator of progressiveness, openness, and youthfulness. They nevertheless experience their skills in some respects as being limited, which they associate with their rather low level of education. Therefore, it is plausible that they limit their interest in technology pragmatically to their own devices.

Due to these motives, the elderly associated with this model have a large number of devices and spend a great deal of time with them. In particular, studying user manuals is considered to be very important to these individuals. This application of time is considered to be the key to improving their technical expertise. The instructions are worked through step-by-step with discipline and perseverance, so as to develop their technological competence systematically. Moreover, training courses are perceived as important chances to deepen their knowledge. It is worth noting that technical problems are experienced as welcome occasions to further develop technical skills.

What are the context conditions for this type? In this sample, the elderly generally have a low-to-medium level of education. Nevertheless, all have professional experience with new technology, which may be the reason for their open-mindedness towards it and their confidence in their ability to master it. Noteworthy is another aspect: Older women were almost forced by the lack of technical expertise of their partner or their boss to acquire a certain degree of technical competence to carry out the necessary tasks successfully. In this respect, the lack of technical competence of an important person in a social environment seems to be a trigger for intensive engagement with technology in others. Nevertheless, their own expertise might still be associated with a positive self-image and pride in the own competence.

Technology as a Passion

Technology has a very special importance for the elderly of this type. It plays a role not only in coping with daily tasks, but above all as a special purpose in life, a passion. Much of their day is devoted to the use of technology. Every technical innovation on the market triggers their desire for learning more about it. At first glance, there exists an analogy to the model *Control over Technology,* because in both types a keen interest in technology itself dominates their motivations. However, there exists one major difference. While the elderly of the model *Control over Technology* desire to increase their technological expertise as a sign of cognitive skills, the people of this type find technology itself at the center of their motivation. They are convinced of their high technical expertise, which is reflected in the fact that they believe they do not need to read user manuals in order to operate new devices.

A main motive for members of this model in the use of technology is, as mentioned above, an intrinsic interest in technical devices. The engagement in technology has as a high priority in their lives as their former employment previously held. Several lines of evidence support the hypothesis that engagement in technology is a replacement for their earlier employment. As previously mentioned, they spend most of the day devoted to technology-based activities, which are strictly separated from their so-called "private life." In addition, these people often receive orders from their social network which require technical expertise, and they fill these orders promptly, dutifully, and almost professionally. Moreover, they did not voluntarily leave their employment, but rather did so out of necessity due to their age. Their employment had become their purpose in life. It seems that the occupation with technology is an attempt at compensation for the lost employment for these elderly.

Each day, these people schedule an intensive engagement with technology. Furthermore, these elderly have a great interest in all technical innovations. They regularly study technical magazines, as well as hold conversations with other technical "experts" about new products. It is self-evident that their households are equipped with a variety of classic and new technology. They perceive every technical problem as a welcome opportunity to be engaged in technology, just as the elderly of the *Control over Technology* model.

How can the context conditions of this model be described? What is most striking about this type is that it is comprised solely of men in this study. They generally have many years of professional experience with sophisticated technical equipment. Accordingly, the intrinsic motivation of technology in itself is already reflected in their choice of career, and may be the trigger that plays a major role in their retirement as well. To this day, technical employment is predominantly held by men. Accordingly, both the interest of technology in itself as well as many years of professional experience with technology could be found especially in men. Therefore, it is reasonable to assume that this type is represented primarily by males.

Construction of the Social Self with Technology

This model shows a proximity to the model *Control over Technology*, as in both models there is a need to ensure one's own competence in using technology. At the same time, there is a clear difference, which justifies constructing a new type: The elderly of this model are able to increase the feeling of their own competence, satisfaction, and pride in dealing with other people through the use of technology. They require social recognition of their technical competence. They make such competence apparent in conversations with friends and relatives, as well as in discussions with technical experts. The inclusion of a variety of devices in daily life is self-evident to them.

The competent use of technology for individuals in this model is closely associated with feelings of social affiliation and social integration. The elderly are convinced they will be excluded from the current and future technological world without sufficient technical skills. In other words, they believe they require this competence to be considered a full-fledged member of society. As one female interviewee says, without these skills, one feels *"to be only half somehow [Irgendwie nur halb zu sein]."* However, the desire for social visibility of one's own high technical expertise has its downsides. These elderly feel pressure caused by continual technical progress. They feel the compulsion to learn constantly in order to keep up-to-date. This is connected with the fear of failure and a fear of being stigmatized as "old" by younger people. Technical expertise for these individuals is associated with youthfulness and progressiveness, traits they seek for themselves.

These motives are expressed in an intense preoccupation with technical devices, especially in terms of time. For example, if a device is to be purchased, these elderly spend a great deal of time carefully weighing the pros and cons of different devices. Not only do they consult the internet to receive aid in the purchase decision, but journals and retailers in specialized stores as well. In addition, technology is an ever-present theme in their discussions with friends and/or their life partners. They also enjoy contact with technical experts such as professors of adult education centers in dealing with technical issues. It is self-evident that the household of these elderly persons include a variety of latest and modern appliances, which are utilized with supreme confidence as a result of their careful research. The elderly see technical problems as welcomed challenges rather than as nursances.

What are the context conditions which promote this interpretative model of technology? Some hints can be derived from the sample and should be examined on their generalizability. All people of this model possess at least an average level of education and have made a wide array of experiences with the use of complex technology as part of their professional activities. In some cases, the intense preoccupation with the new digital technology is intrinsically motivated. In other cases, it was forced due to the requirements of their jobs. Also interesting is a marked gender difference in this model. While the men in this type are predominantly intrinsically motivated, the women experienced mostly an external pressure – such as a boss with a lack of technical expertise – as the trigger for their improvement of technical competence.

Based on many years of professional experience, these elderly persons confidently use technology in their households. Likewise, it seems plausible that they can easily win social recognition with their knowledge, especially in a high-tech world. Because of this, it is easy for them to perceive a sense of belonging and integration in society.

Communication with the Aid of Technology

This type of individual also connects technology with a social need. Technology is considered to be an effective way to achieve permanent access to other people. Accordingly, this interpretative model of technology could be assigned to the Instrumental Dimension. Nevertheless, it is assigned to the Social Dimension because it satisfies the social needs of the elderly in particular. The cell phone, telephone, and email program all play a central role in the daily lives of these elderly. In particular, the cell phone is seen as an opportunity for enlarging and strengthening their own network of relationships and, therefore, is their constant companion.

In this model, the motive for the use of technology is quite clearly the need for permanent and close integration into and interaction with their social network. These elderly always want to be up-to-date when relating to their friends. They are convinced that new media technology allows them an intense contact, which had not been possible in earlier times. Nevertheless, these individuals do not wish to communicate constantly, but rather to be available continuously and to have the opportunity to contact others at any time. Additionally, the cell phone gives them a feeling of affiliation during a "lonely" walk, even without the active use of the device. The spatial absence of the others is transformed by the communication media in a location-independent, interactive presence.

In daily life, this interpretative model of technology is manifested by a competent use of a variety of devices for social interaction. If it is necessary to contact a friend who travels a lot, one female interviewee first sends a text message to determine the location of the friend in order to choose the appropriate mode of communication. Likewise, a male interviewee stated that if he wishes to plan an interaction with his busy daughter, he writes her an email, so she can answer at her convenience. Noteworthy is also the time which is spent during social interaction with the aid of technology. Telephone conversations with friends which cannot be met face-to-face due to geographical distances or (age-related) mobility restrictions often take a great deal of time. Therefore, the telephone flat rate has now become a staple in many households. Concrete reasons for calls are not necessary. Phone calls over "sweet nothings" (Peil 2007: 231) are often used to keep friendships alive. Due to the importance of communication, the elderly put an emphasis on quick solutions to technical issues and were willing to enlist the support of experts in remedying them. Hence, in the perception of this group of elderly, such defects played a subordinate role in their lives.

This interpretative model of technology was represented by both men and women, across all educational levels and age groups. Technical competence, or the lack thereof, seems irrelevant for this model as well. Common to all elderly of this type is that their social network and an intense social interaction play a prominent role in their daily lives.

4. Conclusion

Several trends can be gleaned through the consideration of the motives of the elderly for and against the use of technology. First, the motives associated with the Instrumental Dimension seem to play a prominent role for the majority of the interviewed elderly. Through the use of technology, daily work can be more easily, quickly, and economically managed. As described by Rammert (1988: 192), this could be attributed to the so-called "technological mentality," which has emerged in the course of modern technological development. On this basis, efficiency and improved performance have become an end or a cultural value in themselves. Technology is interpreted primarily as a means to an end. Sackmann and Weymanns' (1994) findings about different technology generations point in this direction as well. According to them, the current interpretation of technology by elderly can be attributed to experiences from their youth, which encourages the idea that technical achievements should primarily facilitate work.

Based on this finding, it can be assumed that factors other than instrumental motives hardly play a role, but it can also be shown that a multitude of other reasons for the use of technology exist. A small proportion of the elderly of this study – which can also be assigned to only one type, the Aesthetic-Expressive Dimension – combines the use of technology with feelings of joy and delight. With beautiful looking devices, they wish to arrange their household aesthetically, coherently, harmoniously, and beautifully. At the same time, these devices serve as a testament to their financial and cultural resources, the latter in the form of good taste. It can be assumed that rather few elderly associate with this model, as it is difficult to develop feelings in the context of instrumental objectivity in dealing with technology as described above (cf. Tully 2003: 148).

In addition, motives for the use of technology can be identified which could be attributed to the Cognitive and the Social Dimensions. Some elderly perceive the use of modern technology as an important opportunity to assure themselves of their own technical skills. Based on this, they consider themselves as open-minded, progressive, mentally flexible, and educated. For others, the use of technology provides an important purpose in life, which is equivalent to that of a profession and represents compensation for their exclusion from employment.

For others, the social visibility of their technical competence plays a major role in their choice to use technology. Through the presentation of their technical expertise, they perceive themselves as belonging to the modern and technological world. Another large group of elderly individuals interacts with technology as an assurance of communicative accessibility. For them, through the use of communication media, the physical absence of their social network is transformed into a location-independent, interactive presence, which is associated with a feeling of affiliation and belonging.

Another finding of the study is that with the use of technology, elderly are usually influenced by several motives simultaneously. For example, some elderly with high technical expertise use technology – in addition to supporting their daily work – as a means of increasing self-confidence, feelings of social affiliation, and social recognition. Avoiding the use of technology leads other elderly to an avoidance of technology-induced loss of control and the visibility of their lack of technical expertise, which contributes to their desire for "invisible" technology.

If it is the aim of innovators to enable older people to live a longer, more independent life through technology, the results of this study must be taken into account during technology development. Not only should technical feasibilities be considered when creating new technology, but the motivations of older people in using that technology should be as well. In order to make this typology of motives for (and against) the use of technology available for technical innovators, it is necessary to develop it further. It is essential to verify the theses on interrelationships between the social, cultural, and structural conditions and the interpretative models of technology with a quantitatively-oriented research design. Based on such studies, groups of people can be identified with specific motives in dealing with technology. In addition, the specific needs of these groups of elderly must be analyzed. The aim of such a collection would be to gather a comprehensive and systematic overview of different groups of older people, which are characterized by specific needs and interpretative models of technology. Technology development should be systematically oriented on these findings.

Acknowledgements

I would like to thank Francesca Fogarty for assisting me with the translation of the text.

References

Ahrens, J. (2009): Going Online, Doing Gender: Alltagspraktiken rund um das Internet in Deutschland und Australien, Bielefeld: Transcript.

Backes, G.M./Clemens, W. (2003): Lebensphase Alter: Eine Einführung in die sozialwissenschaftliche Alternsforschung, Weinheim: Juventa.

Beisenherz, G. (1989): "Computer und Stratifikation". In: Schelhowe, H. (ed.), Frauenwelt – Computerträume, Bremen: Informatik Fachberichte, pp. 93-101.

Bievert, B./Monse, K. (1988): „Technik und Alltag als Interferenzproblem". In: Joerges, B. (ed.), Technik im Alltag, Frankfurt am Main: Suhrkamp, pp. 95-119.

Collmer, S. (1997): Frauen und Männer am Computer: Aspekte geschlechtsspezifischer Technikaneignung, Wiesbaden: DUV.

Giesecke, S. (2003): "Von der Technik- zur Nutzerorientierung – neue Ansätze der Innovationsforschung". In: Giesecke, S. (ed.), Technikakzeptanz durch Nutzerintegration? Beiträge zur Innovations- und Technikanalyse, Teltow: VDI/VDE-Technologiezentrum Informationstechnik, pp. 9-17.

Habermas, J. (1981): Theorie des kommunikativen Handelns, Volume II, Frankfurt am Main: Suhrkamp.

Hargittai, E./Shafer, S. (2006): "Differences in Actual and Perceived Online Skills: The Role of Gender". In: Social Science Quarterly 87/2, pp. 432-448.

Hennen, L. (1992): Technisierung des Alltags. Ein handlungstheoretischer Beitrag zur Theorie technischer Vergesellschaftung, Opladen: Westdeutscher Verlag.

Hörning, K.H. (1988): "Technik im Alltag und die Widersprüche des Alltäglichen". In: Joerges, B. (ed.), Technik im Alltag. Frankfurt am Main: Suhrkamp, pp. 51-94.

Hörning, Karl H. (1989): "Vom Umgang mit Dingen. Eine techniksoziologische Zuspitzung". In: Weingart, P. (ed.), Technik als sozialer Prozeß, Frankfurt am Main: Suhrkamp, pp. 90-127.

Hopf, C. (1991): "Qualitative Interviews in der Sozialforschung: Ein Überblick". In: Uwe Flick et al. (eds.), Handbuch qualitative Sozialforschung:

Grundlagen, Konzepte, Methoden und Anwendungen, München: Psychologie Verlagsunion, pp. 177-182.

Joerges, Bernward (1988): "Gerätetechnik und Alltagshandeln". In: Ders. (ed.), Technik im Alltag, Frankfurt am Main: Suhrkamp, pp. 20-50.

Kaufmann, Jean Claude (1999): Das verstehende Interview: Theorie und Praxis, Konstanz: Universitätsverlag.

Kelle, U./Kluge, S. (1999): Vom Einzelfall zum Typus: Fallvergleich und Fallkontrastierung in der qualitativen Sozialforschung, Opladen: Leske + Budrich.

Kreibich, R. (2004): "Selbstständigkeit im Alter: Neue Dienstleistungen, neue Technik, neue Arbeit". In: ArbeitsBericht Nr. 3/2004 des IZT (Institut für Zukunftsstudien und Technologiebewertung). Vortrag auf dem Workshop des Bundesministeriums für Bildung und Forschung. Leitvision "Selbstständigkeit im Alter – Dienstleistungen und Technologien." Bonn.

Peil, C. (2007): "Keitai-Kommunikation: Mobiler Medienalltag in Japan". In: Röser, J. (ed.), MedienAlltag: Domestizierungsprozesse alter und neuer Medien, Wiesbaden: VS, pp. 223-233.

Pelizäus-Hoffmeister, H. (2013): Zur Bedeutung von Technik im Alltag Älterer, Wiesbaden: VS Verlag.

Rammert, W. (1988): "Technisierung im Alltag. Theoriestücke für eine soziologische Perspektive". In: Joerges, B. (ed.), Technik im Alltag, Frankfurt am Main: Suhrkamp, pp. 165-208.

Rammert, W. (2007): Technik – Handeln –Wissen: Zu einer pragmatischen Technik- und Sozialtheorie, Wiesbaden: VS Verlag.

Sackmann, R./Weymann, A. (1994): Die Technisierung des Alltags. Generationen und technische Innovationen, Frankfurt am Main: Campus.

Saup, W. (1993): Alter und Umwelt. Eine Einführung in die Ökologische Gerontologie. Stuttgart: Kohlhammer.

Saup, W./Reichert, M. (1999): "Die Kreise werden enger: Wohnen und Alltag im Alter". In: Niederfranke, A., Naegele, G., Frahm, E. (eds.), Funkkolleg Altern 2. Lebenslagen und Lebenswelten, soziale Sicherung und Altenpolitik, Opladen: Westdeutscher Verlag, pp. 245-286.

Schulze, G. (1992): Die Erlebnisgesellschaft. Kultursoziologie der Gegenwart, Frankfurt am Main: Campus.

Strauss, A.L./Corbin, J. (1996): Grounded Theory: Grundlagen Qualitativer Sozialforschung, Weinheim: Fink.

Technische Universität Berlin (2011): Nutzerabhängige Innovationsbarrieren im Bereich altersgerechter Assistenzsysteme. 1. Studie im Rahmen der AAL-Begleitforschung des Bundesministeriums für Bildung und Forschung. www.aal-deutschland.de/deutschland/dokumente/Abschlussbericht AAL-Nutzerstudie_Final.pdf (last access: 22.07.13).

Tully, C. (2003): Mensch – Maschine – Megabyte: Technik in der Alltagskultur. Eine sozialwissenschaftliche Hinführung. Opladen: Leske + Budrich.

Wengenroth, U. (2001): "Vom Innovationssystem zur Innovationskultur". In: Abele, J., Barkleit, G., Hänseroth, T. (eds.), Innovationskulturen und Fortschrittserwartungen im geteilten Deutschland, Köln: Böhlau, pp. 23-32.

World Health Organization (WHO) (2002): Aktiv altern: Rahmenbedingungen und Vorschläge für politisches Handeln, http://whqlibdoc.who.int/hq/2002/WHO_NMH_NPH_02.8_ger.pdf (last access: 07.06.13).

Witzel, A. (2000): "Das problemzentrierte Interview". In: Forum Qualitative Sozialforschung/Forum Qualitative Social Research 1/1, Art. 22, http://nbn-resolving.de/urn:nbn:de:0114-fqs0001228 (last access: 09.10.13).

Woolgar, S. (1991): "Configuring the User: the Case of Usability Trials". In: Law, J. (ed.), A Sociology of Monsters: Essays on Power, Technology and Domination, London: Routledge, pp. 57-99.

An Exploration of Mobile Telephony Non-use among Older People

MIREIA FERNÁNDEZ-ARDÈVOL

1. INTRODUCTION

What if I get my first mobile phone when I'm 75 years old? What if I decide to give it up at that age? Older people are commonly qualified either as later adopters or as the less innovative users of mobile communication (for instance, see Karnowski/von Pape/Wirth 2008). Challenges and difficulties older people face seem to justify why they are not interested in using mobile devices: therefore, a significant number of research is focused on usability. Studies that focus on mobile phone use among older people describe the prevalence of voice communication (see Kurniawan/Mahmud/Nugroho 2006; Ling 2008; Lenhart 2010) and how pressures from relatives can foster use (Ling 2008). Yet mobile phones need to be useful, social –because they allow communication among individuals, that is engagement– and enjoyable to be adopted (Conci/Pianesi/Zancanaro 2009). They usually constitute an extra layer of communication, used in addition to and complementing the landline –at least in developed countries (Fernández-Ardèvol/Arroyo 2012).

In this paper I focus on mobile phone non-use and the meaning this situation has in the case of older people, that is, for individuals aged 60 years old and over (60+). Given the pervasiveness of mobile phones, research questions are: What are the motivations older people have for not using mobile phones? How does this decision shape, if it does, mediated connectivity? What are the perceived effects, positive and negative, of non-ownership? While responding to these questions would be relevant for any age group, I focus on older population because they constitute the age group that has been less studied in this field.

I draw my discussion on qualitative research conducted in five cities between 2010 and 2013: selected metropolitan areas bring diversity both in socioeconomic terms and regarding telecommunication markets. The cities are Barcelona (Catalonia, Spain); Los Angeles (California, USA); Toronto (Ontario, Canada); Montevideo (Uruguay) and Lima (Peru). We held interviews with individuals 60+ around mobile communication.[1] The paper will analyze common patterns of mobile phone non-use among this subsample of participants. Up to 39 participants out of 147 identified themselves as mobile phone non-users at the moment of our conversation, describing an array of situations when explaining their relationship with mobile devices.

Main conclusions, first of all, have to do with heterogeneity: the simple label "non user" hides a variety of situations in the case of older people. The analysis shows that motivations for non-use are not only the challenges and difficulties I mention above: beyond the commonly accepted reasons, non-use can be a personal decision, a kind of pushback strategy that shows the older person's agency. Besides, interviews demonstrated that being a mobile phone user or not can follow a dynamic path; individuals can, and do, change their minds regarding whether to use a mobile phone or not.

Results are relevant *per se* as there is a need of in-depth empirical evidence on the relationship older people have with information and communication technologies (ICT) in general and with mobile technologies in particular. In addition, as the path of innovations is not going to decrease, current results will also help to better understand the way future new information and communication technologies are adopted, rejected or ignored by older people.

2. Conceptual Framework

Mobile phones are not isolated technologies but part of a landscape that includes other communication media –for instance, landline phones. Media, in this sense, can be approached as an 'ecology' or assemblage of interconnected technical elements (Latour 2005). Individuals would usually make a choice among devices when deciding whether, or not, to incorporate mobile

---

1 | Case studies relayed in semi-structured interviews except for the case study in Lima, where the author also conducted two discussion groups. Length of conversations varied from country to country, with an average of around one hour conversation. Case studies are part of a research line on older people and mobile communication of the research group Mobile Technologies and (G)Local Challenges (IN3, Open University of Catalonia).

phones into their everyday life (Hashizume/Kurosu/Kaneko 2008). Whatever their decisions are, every individual would identify a specific set of channels in their everyday life, their personal set of communication channels (PSCC). Individual attitudes and aptitudes, as well as personal interests and socially imposed interests or pressures might shape the PSCC (Fernández-Ardèvol/Arroyo 2012). An important aspect to take into account is "the context in which communication processes occur" (Foth/Hearn 2007: 9), as it helps to understand "the impacts and possibilities of a particular medium, and how [mediated] communications fit into the other things that people are doing" (Tacchi/Slater/Hearn 2003: 15). Still it becomes relevant to consider the specific goals (Ball-Rokeach 1985) allowed by different forms of mediated communication, and whether the different communication media available to each individual help them reaching goals in each context. In turn, goals of communication evolve with time because personal communication evolves with age (Charness/Parks/Sabel 2001).

Different typologies of use arise from empirical evidence and in response to different research questions. For instance, Brandtzæg, Heim and Karahasanović (2011) distinguish non-users, sporadic users, instrumental users, entertainment users and advanced users of the Internet. With regard to mobile phones, Oulasvirta, Wahlström and Ericsson (2011) characterize non-users, novice users, casual users and professional users. Policy makers interested in closing digital divides are focused on increasing use: in general, available indicators assume that the higher the level of adoption, the better. They often focus on skills and context conditions to reduce non-use; indeed, affordances of use (Galperin/Mariscal 2007) give more information than the mere dichotomous variable use / non-use. Affordances of use refer to the capacities individuals have to effectively conduct the activities they would like to conduct with specific ICTs. Therefore, in terms of public policies, the analysis of use mainly focuses on individuals' digital activities and their frequency, as these are indicators of use capacity –see, for instance, the Digital Agenda for Europe (European Commission 2015).

Given these assumptions, it is common to find works where non-use is approached as a problem to be solved, but "defining people only as producers or as users of technologies confirms the technocratic vision of the centrality and normativity of technology. Users of technology also need to be seen in relation to another, even less visible group – namely non-users." (Wyatt/Thomas/Terranova 2002: 25) Following Wyatt/Thomas/Terranova (2002), I avoid approaching non-use as a deficit. Non-use can be either voluntary – as it is for resisters and rejecters – or involuntary – as it is for excluded and expelled (Wyatt/Thomas/Terranova 2002). Similarly, it can be volitional – individuals not interested or those who already gave up using – and non-volitional –individuals who, for different reasons, face lack of knowledge or lack of access to use these communication technologies (Brady/Thies/Cutrell 2014).

When individuals reach saturation, "situational non-use" can appear. That is, established users of online social technologies would occasionally stop using them; that situation can even turn into a permanent drop-out (Leavitt 2014). While situational non-use often refers to online advanced services, the concept can also apply to basic electronic tools such as mobile phones. Perceived usefulness, which shapes adoption, does not exist as a stable category (Chirumamilla 2014): this is one of the reasons justifying changes over time. In this sense, Fausset et al. (2013) approach older adults' perception of technology by using a behavioral change model – the stages of change model. As for this approach, there is always a possibility to stop using a given technology at any point in the process of adoption and appropriation. In line with this, Wyatt (2014) claims there is a need for a new conceptual framework that must include both users and non-users as well as practices of use and non-use, as users of digital technologies need to be conceptualized along a continuum with degrees and types of involvement that may change depending on personal characteristics and circumstances.

It is commonly accepted that older age starts with retirement. Regardless of the specific age (60 or 65 years old), a distinctive characteristic of the category "elderly people" is its high heterogeneity as "lives fan out with time" (Neugarten 1996). Not only educational level, but also life experience and life trajectories create diversity. This should be taken into account when approaching this wide age group which, indeed, gathers different generations under the same label. Neugarten suggested the distinction between young-old and old-old as a "gross way of acknowledging some of the enormous diversity among older persons" (Neugarten 1996: 48) that overcomes the limitations of classifications strictly based on age. The young-old refers to retired individuals, healthy and integrated members of their families and communities. The old-old refers to persons who need a range of health and social supportive/restorative services.

Despite their diversity, industry research often considers older people as a homogeneous group, what Sawchuk and Crow (2011) call the grey zone. This approach can also be found in some academic publications – as is claimed by Loos, Haddon and Mante-Meijer (2012). Yet, some studies on ICT and ageing do acknowledge such heterogeneity: two examples serve as illustration. The first one, which focuses on the old-old, is a study on the information behaviors in the fourth age aimed to conceptualize information literacy (Williamson/ Asla 2009). Authors conclude that, taking into account the limited research on the area, common assumptions on information behaviors in the fourth age are not well founded. The second one focuses mostly on the young-old: the paper claims for heterogeneity in age by studying ICT use by people between 55 and 65 years old – that is, in transition to retirement – as this age group is rarely considered in studies. A focal point is that late-midlife people are challenging "the assumption that they find it difficult to embrace new technologies" (Salovaara et al. 2010: 803). Both examples criticize assumptions on how older people use ICTs. These assumptions are sometimes built upon age norms that construct older people as limited individuals, as "object of others' actions, reducing their prospects of being seen as actors entitled to making their own decisions" (Jolanki 2009: 215). Linked to this idea of agency, it would be a mistake to assume that any older individual not using mobile phones will be, by default, more isolated or will have fewer opportunities to develop their personal interests. Such causal relationship would respond to a non-critical techno-deterministic approach (as argued by Pinch/Bijker 1987).

3. CASE STUDIES

I draw my discussion on qualitative research conducted in five cities between autumn 2010 and autumn 2013: the cities are Barcelona (Catalonia, Spain); Los Angeles (California, USA); Toronto (Ontario, Canada); Montevideo (Uruguay) and Lima (Peru). In each city we conducted an independent case study that followed a common methodological and conceptual framework to allow for a thorough insight into a complex phenomenon from a qualitative perspective and favors the search of common trends that do not depend on contextual settings (Yin 2003). Selected metropolitan areas brought diversity in terms of culture, socioeconomics and telecommunication landscapes. Besides, the generational diversity of participants was guaranteed by establishing only a lower boundary in age – 60 year old – but no upper boundary. A flexible, interactive research design was taken to incorporate the specific circumstances in which the research was carried out (Maxwell 2005). In the paragraphs that follow I will describe the most relevant characteristics of the case studies discussed in this paper.

The goal of the original research project was to understand the relationship older people have with and through mobile telephony. Yet, mobile ownership was not a selection criterion in order to allow the gathering of relevant information for a better understanding of the processes of rejection and acceptance of this technology. Particular attention was paid to the different communication media individuals use in their everyday life, as these allowed us to achieve a better understanding of the specific role of mobile communication within the communicative ecology of each individual.

Case studies relayed in semi-structured interviews except for the case study in Lima, where the author also conducted two discussion groups. A short questionnaire and personal notes taken after each interview complemented information gathering. First of all, every semi-structured interview and discussion group followed a flexible outline and was conducted as a relaxed conversation. Conversations revolved around the communication media individuals use (mainly landline, mobile phone and Internet); the people and institutions they communicate with in their everyday life; and the common uses of communication channels, with a specific focus on mobile phones – if used. The interviews also discussed motivations, opinions and personal experiences regarding the decision to have a mobile phone or not. First, conversations were voice recorded for further text analysis and research ethics were adapted to each country context. Respondents were at all times informed that they could skip any question they did not feel comfortable with and that they could end the conversation whenever they wanted. Secondly, the questionnaire helped to gather structured socioeconomic information. Finally, notes taken after the interview aimed at incorporating non oral information of the whole interaction; for instance, relevant pieces of information that were gathered when the voice recorder was off and, in general, researcher reflections before, during and after the interview.

The number of participants totaled 147, with ages ranging between 60 and 98 years old. The sample was composed by more women (103) than men (44), a common trend in all the case studies. In order to recruit participants, we relied on informants and on snowball sampling. In the metropolitan area of Barcelona we interviewed 53 participants aged between 60 and 96 years old between October 2010 and March 2011. Participants lived in different areas of the metropolitan area, so we accessed individuals of different socioeconomic conditions. Among them, 12 participants were living in retirement homes. In Los Angeles, I conducted fieldwork in autumn 2011: I accessed 20 older people, most part of them volunteers in an organization, with ages between 61 and 92. While most of them lived in middle-class neighborhoods, some also lived in more modest areas. In Toronto, we conducted interviews in May 2012 with 22 participants during 2 days in a high-end retirement home. This group constituted the most aged section in our sample, with individuals ranging from 75 to 98 years old. Montevideo, where I interviewed 15 participants in June 2012, gathered the younger group with ages ranging from 64 to 75. Finally, in Lima I talked to 37 older individuals in September 2013, their ages ranging from 61 to 98. While in Montevideo I defined no condition on the income level of participants and I accessed middle- and upper-class population, in Lima I focused both in middle- and upper-income neighborhoods, but also in a low-income district to allow for a more thorough analysis.

In order to answer the specific research questions proposed in this paper, I conducted thematic analysis of the conversations and the personal notes of participants who identified themselves as non-users. My focus stayed in the analysis of what was said during the interviews, which was usually a projection, that is, a narrative of what happens in each individual's life (Kvale 1996). Among the 147 participants, 39 identified themselves as mobile phone non-users at the moment of the conversation (26 women and 13 men). I will develop those cases which I consider better illustrate the main ideas found in my analysis, regardless of the case study they belong to.

4. RESULTS

All the participants in the examples had access to a landline, unless otherwise stated, and used it regularly in their everyday life communications. Those in the labor market were regular users of computers and the Internet, while some retired participants did also go online. No real names are used to preserve anonymity.

Paco (84 years old) tried to use a mobile, but he already broke two devices. He then decided he did not want to have one due to those discouraging negative experiences:

I have it in my pocket and it's like I forget it... I don't, I don't hear it. [...] Then I make a real mess to call... [...] I don't have memory for that. (Paco, 84, Barcelona)*[2]

---

2 | Legend: * Own translation, from Catalan or Spanish; ** English in the original.

Andreu (88) already adopted a mobile phone and got used to it, but he suffered a stroke that affected his dexterity. He then stopped using the device but,

I keep it [the mobile phone] just in case one day I need to use that [small] part [the SIM card] in another handset... [or] I need the data stored on it (Andreu, 88, Barcelona)*

Still, the mobile can become useless for reasons not related to usability, as Ann and Consuelo explained. Ann (63) and her husband decided to buy a mobile phone in 2005:

We both bought cell phones when we figured that we should probably have it for emergency purposes. And then... And also it was cheaper if you bought two. You had two for the price of one. [...] So it was... it made sense that I had this. So I just didn't use [it]. So I kept it in the glove compartment of my car. "So this is perfect. In case of an emergency I'll have the cell phone." It just was pointless. I think it's still in the... It's still in my glove compartment. (Ann, 63, Los Angeles)**

In the few occasions Ann needs a mobile phone, for instance when picking her daughter up from the airport, she uses her husband's device. Her friends sometimes tease her for not having a mobile phone, but this is not a big deal for her.

Consuelo (67) also decided to buy a mobile phone in 1993. At that time, mobile calls were significantly more expensive than landline communications:

I did it for my sons... because of them and because I didn't want to stay that many hours at home, as I was coordinator of a friendship group. The more I was behind the landline the higher the possibilities for people to join the activities of the center. [...] Then I thought: "Look, a mobile phone will give me more freedom", but at that time almost anyone had mobile phones and I was a little ahead [...]. And it turned to be useless, because everybody called me on the mobile to ask for my landline number, and then I thought "I don't want to pay that bill when people want the landline", and I gave up." (Consuelo, 67, Barcelona)*

At the moment of the interview Consuelo was not using any mobile phone. A friend gave her a device as a present, but she does not feel she needs it. Yet she is clear about priorities: if one of her sons asked her, she would use the mobile.

Wilma (67), who never got used to the mobile phone she received as a present from her son, deployed an interesting strategy for being always reachable. She does not feel she needs the mobile, but it is important for her not to worry her sons:

I put papers, I tell you, I write in big white papers: "I'm leaving, it's 19h, I'm at the prayer group", [and] there's a phonebook with all the numbers (Wilma, 67, Montevideo)*

Clementine (76) also received a mobile phone as a present but she rejected it immediately. She does not surrender to her family pressures:

Yes, they do [pressure], every day they see me. "You should have a cell phone, something could happen to you. [...]" And I tell them, "There's a phone on the corner, there is phones in the store, there are phones everywhere. There is people around that have a phone that maybe they would call them for me." (Clementine, 76, Los Angeles)**

Another strong rejecter is Laura (63): she manages at her job without a car and without a mobile phone. She agrees she is not an "average" person as she is against consumerism, and explains her priorities clearly:

I look for my own [personal] times. But when you have a mobile phone, before you realize it they are already here; everyone is invading your time. And you must be there, in tune with what the other says. Well… on this I'm really resistant, ok? (Laura, 63, Barcelona)*

Esther (87) also considered the idea of having a mobile phone, but has a straightforward opinion:

"I've decided I don't need one [mobile phone]. (…) And I don't want one. They seem to be getting more and more complicated. (Esther, 87, Toronto)**

In contrast, César (68) never considered buying a mobile phone. However, if he were given one as a present he would be happy with it:

"I won't give [the mobile phone] back. I will keep it. (César, 68, Montevideo)*

This was somehow surprising for his son, who eventually explained that her sister and he did not expect Carlos using it that much. Instead, they would expect him to stop using the device once the battery drains, as he would not know what to do then.

Blanca (98) is used to have domestic workers at home. She is living on her own with two workers who take care of domestic duties and, now she is more dependent, of her as well. At this point in her life, many more tasks are done for her than in the past; for instance, she never goes out on her own. Thus, she does not need a mobile phone because

When I go out I'm always with the maid. They [anyone who would like to reach her] can call her. (Blanca, 98, Lima)*

Stephan (78) used to be an international, freelance professional and is now retired. When asked, he said he is not against the mobile phone; it was just that "I don't need it". However, he might changing his decision in the future:

If something came up and I got interested in doing something again that required it [the mobile phone]. [...] Not just a bunch of crappy, goofy conversation. If legitimate function was necessary, fine. (Stephan, 78, Los Angeles)**

In the retirement home in Toronto, where landlines are installed in residents' rooms the moment they move to the premise, most part of non-users told us they never considered having a mobile phone. Particularly, John (98) never considered having one. He is a wealthy, self-made man who is proud of himself. Money is not an issue for him:

"I haven't found any use for it [the mobile phone]. [...] The lifestyle...I mean, I'm not hopping all over and... if I was going around with...and my wife was alive and something... and you were going downtown, I want to phone her or something, I would have one. [...] So [...] I can't be bothered with it. And I don't miss it. But I haven't...if I needed it I wouldn't hesitate to get it." (John, 98, Toronto)**

Cora (76) lives in a low-income district. She would be happy to receive a mobile phone as a present for a second time and "learn how to use it again". She explains she "enjoyed" having the mobile phone she received as a Christmas present in 2009, but her device was stolen after few months and, after the experience

They [her children] won't buy me another one. Not anymore (Cora, 76, Lima)*

Catia (88) lives in the same district. She was a hard-working single mother, currently living with two of her adult children. They have a pay-as-you-go landline phone with no airtime, so her children cannot call her when they are late. When asked, she says she would like to have a mobile phone and that this would be the only phone for her:

But I should practice to use it, [because] once I had one [but]... it flew away. (Catia, 88, Lima)*

She refers to a used mobile phone her daughter borrowed her so that she could call Catia during the day: a caregiver broke it and the daughter threw it away but never replaced it.

Finally, Evelyn (90) would also like to have a mobile phone. She explains she is legally blind and already has a phone home with big buttons so she can operate the device properly. But the landline is not enough, as she has a pretty active life:

[...] they're making larger buttons but I just haven't been able to get one yet. [...] Yes. I would like to have one with larger letters. [...] And then I could take it with me. [...] because I go to classes and things and I need something to.... [...]. And I go to Braille [classes] and most of the people there have the cell phones. They teach you how to use them. (Evelyn, 90, Los Angeles)**

5. DISCUSSION

The first evidence is that participants identifying themselves as non-users are both young-old and older-old. Some of them never had a mobile phone (Laura, 63 years old; Clementine, 76; César, 68; Blanca, 98; Stephan, 78; John 98; and Cora, 76), while others had some experience with a device (Paco, 84; Andreu, 88; Ann, 63; Consuelo, 67; Wilma, 67; Cora, 67; and Catia 88). Yet a closer look to these wide categories is necessary for a more detailed picture of the motivations and conditions of non-use in the case of older people. I organized the results under four headlines corresponding to categories, although these are not mutually exclusive.

A Present I Keep in a Drawer

It is common to give a mobile phone as a present. Older people are not an exception and it is a way for relatives to foster adoption and it can also act as a gentle form of pressure (Ling 2008), although this strategy does not always work. This is the case of Paco (84), who tried to get used to the mobile but it proved not to be useful, neither socially engaging nor entertaining –or enjoyable (Conci et al. 2009). This is also the case of Clementine (76), who did not even try out her present and rejected using it. Her case reveals she had agency – or free will – to face her family pressures; her rejection demonstrated she was not the object of others' decisions (Jolanki 2009).

I don't Like it!

Laura (63) and Esther (87) also demonstrated their agency when they decided not to have a mobile phone. Even though both participants are under the label 60+, Laura would qualify as a young-old and Esther as an old-old (Neugarten 1996). Their reasons for rejection seem to be shaped by their vital circumstances. Esther would expect unnecessary complications if she had a mobile phone; in contrast, Laura does not discuss usability issues. I argue that, in this case, as mobile technologies are pervasive and have a high symbolic value, rejecting the mobile phone also creates identity. In both cases, part of their identity is built around rejection of consumerist behaviors, which would include using a mobile phone, and that such behavior is related to ideology (Selwyn 2003).

I don't Feel I Need it

In contrast with participants who express a clear position against mobile phones, most non-users argued they already have other channels for mediated communication (the landline, for instance), so they do not need another layer of communication. Blanca (98) would constitute an extreme case: she does not need a mobile phone of her own, as her domestic workers would take care of her mediated communications any moment of the day and they already have one. This does not seem to affect her agency, her free will, as she has got domestic workers 24h a day during her whole life.

Participants who do not feel the need for a mobile phone describe different attitudes towards the idea of having such a device. First, John (98) explains that in other moments of his life he could have used a mobile phone, but now it does not fit in his lifestyle. Conversely, Stephen (78) would consider having one in case he would like to join an activity that justified using a mobile. Their age difference – 20 years – would be related to their dissimilar perspectives. Second, César (68) never considered having a mobile phone. The initial unawareness he expressed was nuanced by his predisposition to use it if he would be given one. Interestingly, his case illustrates how "preconceptions about [older people] needs and capabilities" (Östlund 2004: 46) operate. His son was surprised when Carlos expressed that attitude: Carlos' children, in fact, would expect him not to use it as he would never understand the device.

Finally, initial reluctance to mobile phones among older individuals can turn into acceptance if the service meets their personal needs (Ling 2008). However, for Ann (63) and Consuelo (67) the experience was the opposite: they tried to use a mobile phone but it turned out to be useless; this is why they did not feel they needed it. They constitute an example of technology pushback, understood as a negotiation that calls into question the norms surrounding the use of digital tools (Young et al. 2014). In the same way, Wilma (67) does not need to have a mobile phone to achieve her goals of communication (Ball-Rokeach 1985), as she built a whole strategy for being reachable when out of home that involved the cooperation of her personal network. While I do not have enough evidence about the perceptions of her friends and relatives, this case would qualify as a "purposeful non-adopter [that can be] unflatteringly described as parasitic." (Katz 2008: 435)

I'd Like to Have one but...

Up to now, the participants made a choice (Hashizume/Kurosu/Kaneko 2008) and discarded the mobile phone. We already saw that some assume – implicitly or not – their decision is straightforward, while others would be open to change it. However, there are participants who look forward to having a mobile phone; this is the case of Evelyn (90), who was not yet able to get a device adapted to her visual impairment. Her attitude seems different to that of Andreu (88), who just stopped using the device after he suffered a stroke. One reason could be that Evelyn joins in a lot of activities out of the home, while Andreu is now comparatively much less active.

Participants living in a low-income district in Lima faced economic restrictions. For Cora (76) the mobile phone would be a complement to the landline, but for Catia (88) it would constitute her only available phone when she is alone at home. Interestingly, the three women seem to have lot of expectations about mobile phones, as they mention they would be happy to learn how to use the device. Their attitude is the opposite to that of Esther (87), John (98) or Paco (84), who refer to difficulties in either their actual or their expected experience with the mobile phone to justify their non-adoption position.

Research Questions

My first research question looked at the motivations of older people for not using mobile phones. It is possible to divide the participants' motivations for non-use in two categories: the first one would include resistance (I don't like it), uselessness (I don't need it) and unawareness (I never considered it). Resistance and uselessness might be a consequence of challenges and usability issues but, interestingly, they might also be an ideological response of older individuals. Indeed, they can even appear after some experience with using a mobile phone. The second category of motivations is related to economic or physical restrictions: as opposed to agency, these are limiting factors that remove the ability to decide from the older person. While there is not an exact correspondence between these two categories, they are similar to voluntary / involuntary (Wyatt/Thomas/Terranova 2002) and volitional / non-volitional non-use (Brady/Thies/Cutrell 2014).

The second research question looked at how mediated communication is shaped by mobile non-use while the last research question looked at perceived effects, positive or negative, of mobile non-ownership. Not having a mobile phone can become a problem for others, and specific strategies can appear to cope with expected moments of disconnection. However, when non-use is a consequence of a personal decision, participants tend to report neutral or positive effects. If there is any negative effect, they explain they can cope with it. When non use is the consequence of a limiting factor, such as an economic or a physical restriction, non-ownership is evaluated in negative terms and the mobile phone appears to be a desired object. The mobile phone is seen as a tool that would improve their needs for mediated communication while being outdoors. For the rest of everyday situations, as long as they had access to a landline, they did not report any limitations.

6. CONCLUSIONS

This paper explored the motivations and experiences around mobile non-use as reported by older individuals who, regardless of whether they had a mobile phone kept in a drawer or not, defined themselves as mobile phone non-users.

My main conclusions, first of all, have to do with *heterogeneity*. The simple label "non user" hides a variety of situations in the case of older people. Some seniors reject mobile phone adoption and fight against relatives' or friends' pressures, thus reinforcing their identity with this decision, while others assert that mobile communication has no place in their universe. The category also gathers individuals who never owned a mobile phone, as well as other older individuals who used to have a mobile device. In terms of heterogeneity, this paper also contributes with empirical evidence to build age nuances regarding the relationship older people have with mobile telephony.

Secondly, *the availability of other communication channels is relevant*. In general, when there is a fixed line, the mobile phone may play a peripheral position within the personal system of communication channels in the older person's life. Seniors for whom mobile devices are complementary can easily manage without them: this appears to be even more obvious in the case of older individuals who decided to stop using the mobile phone they already had. Indeed, disconnection only appears when there is not any available landline.

Thirdly, it becomes necessary to open the dichotomous classification "user / non-user" to have a better picture of the relationship individuals have with and through mobile phones regardless of their age. Particularly, *(non-) use should be approached as a dynamic category* as, by definition, (non-) use is not a permanent state. Our lives evolve; our interests and our goals of communication evolve; and the digital devices available in the market also do. Therefore, every individual at any age is continuously making decisions regarding adoption, non-adoption, or rejection of digital forms of communication. Gathered evidence shows that older users are not an exception.

Fourth, while I focus in the general trends arising from the case studies, poverty arises as a context-specific issue in the low-income neighborhood I accessed in Lima. When purchase capacity is limited, having a mobile phone becomes a luxury for older people. Non-voluntary non-use is an indicator of lack of agency – or limited free will. In these cases non-users would express a desire to have a mobile phone due to the social relevance of the device.

Finally, when there is voluntary non-use, decisions are related to agency. Particularly, non-use is motivated by resistance and uselessness, but also by unawareness, and can be understood as a form of agency of older individuals. We should approach decisions on non-use at any age as seen from the subjective perspective of individuals. In this sense, refusing to use mobile phones can be an ideological decision or a way to cope with personal situations that change as we age.

Due to the growing pace of innovations, the results obtained in this study will also help to better understand the way future new information and communication technologies will be adopted, rejected or ignored by older people. Some evidences – as mobile phones kept in drawers, ignored, and regarded as good-for-nothing – bring a suggestive starting point for future research in this arena. For instance, it would be interesting to look to the intensity individuals have with mobile phones from a longitudinal perspective: When do a mobile phone gets into, or jumps out, from a drawer? Or What are the circumstances that change the relationship older individuals have with mobile phones?

Acknowledgments

The author would like to thank the participants in each case study, as well as the facilitators. Fieldwork in Toronto was conducted as an activity of the research network "Ageing Communication Media" (http://a-c-m.ca; http://actproject.ca/). Part of the research received support from SSHRC Canada (ref. code: 895-2013-1018). Usual disclaimer applies.

References

Ball-Rokeach, S.J. (1985): "The Origins of Individual Media-system Dependency: A Sociological Framework." In: Communication Research 12/4, pp. 485–510.

Brady, E./Thies, W./Cutrell, E. (2014): "No Access, no Knowledge, or no Interest? Examining Use and Non-Use of Assistive Technologies." In: Refusing, Limiting, Departing, In: CHI 2014 Workshop Considering Why We Should Study Technology Non-use, Toronto. http://nonuse.jedbrubaker.com/wp-content/uploads/2014/03/non-use-assistive-tech.pdf (last accessed 02/07/2015).

Brandtzæg, P.B./Heim, J./Karahasanović, A. (2011): "Understanding the New Digital Divide—A Typology of Internet Users in Europe." In: International Journal of Human-Computer Studies 69/3, pp. 123–138.

Charness, N./Parks, D./Sabel, B. (eds.) (2001): Communication, Technology and Aging: Opportunities and Challenges for the Future, New York: Springer.

Chirumamilla, P. (2014): "The Unused and the Unusable: Repair, Rejection, and Obsolescence." In: Refusing, Limiting, Departing, In: CHI 2014 Workshop Considering Why We Should Study Technology Non-Use, Toronto. http://nonuse.jedbrubaker.com/wp-content/uploads/2014/03/2014_position_paper.pdf (last accessed 02/07/2015).

Conci, M./Pianesi, F./Zancanaro, M. (2009): "Useful, Social and Enjoyable: Mobile Phone Adoption by Older People." In: Gross, T./Gulliksen J./Kotzé P./Oestreicher, L./Palanque, P./Prates, R.O./Winckler, M. (eds.), Human-Computer Interaction – INTERACT 2009. Lecture Notes in Computer Science Volume 5726, pp. 63-76.

European Commission. (2015): Digital agenda scoreboard. http://ec.europa.eu/digital-agenda/en/digital-agenda-scoreboard (last accessed: 03/07/2015).

Fausset, C.B./Harley, L./Farmer, S./Fain, B. (2013): "Older Adults' Perceptions and Use of Technology: A Novel Approach." In: Stephanidis, C./Antona, M. (eds.), Universal Access in Human-Computer Interaction. User and Context Diversity. Lecture Notes in Computer Science Volume 8010, pp. 51-58.

Fernández-Ardèvol, M./Arroyo, L. (2012): "Mobile Telephony and Older People: Exploring Use and Rejection." In: Interactions: Studies in Communication & Culture, 3/1, pp. 9–24.

Foth, M./Hearn, G.N. (2007): "Networked Individualism of Urban Residents: Discovering the Communicative Ecology in Inner-City Apartment Buildings." In: Information, Communication & Society 10/5, pp. 749–772.

Galperin, H./Mariscal, J. (eds.) (2007): Digital Poverty: Latin American and Caribbean Perspectives, Bourton-on-Dunsmore: Practical Action Publishing.

Hashizume, A./Kurosu, M./Kaneko, T. (2008): "The Choice of Communication Media and the Use of Mobile Phone Among Senior Users and Young Users." In: Lee, S./Choo, H./Ha, S./ Shin, I.C. (eds.), Computer-Human Interaction. Lecture Notes in Computer Science Volume 5068, 2008, pp 427-436.

Jolanki, O.H. (2009): "Agency in Talk about Old Age and Health." In: Journal of Aging Studies 23/4, pp. 215–226.

Karnowski, V./von Pape, T./Wirth, W. (2008): "After the Digital Divide? An Appropriation-Perspective on the Generational Mobile Phone Divide." In: Hartmann, M./Rössler, P./Höflich, J.R. (eds.), After the Mobile Phone? Social Changes and the Development of Mobile Communication, Berlin: Frank & Timme, 4, pp. 185–202.

Katz, J.E. (2008): "Mainstreamed Mobiles in Daily Life: Perspectives and Prospects." In: Katz, J.E. (ed.), Handbook of Mobile Communication Studies, Cambridge: MIT Press, pp. 433–445.

Kurniawan, S./Mahmud, M./Nugroho, Y. (2006): "A Study of the Use of Mobile Phones by Older Persons." In: CHI 2006 Extended Abstracts on Human Factors in Computing Systems, New York: ACM, pp. 989–994.

Kvale, S. (1996): InterViews: An Introduction to Qualitative Research Interviewing, Thousand Oaks: Sage Publications.

Latour, B. (2005): Reassembling the Social: An Introduction to Actor-Network-Theory. Oxford: Oxford University Press.

Leavitt, A. (2014): "When the User Disappears: Situational Non-Use of Social Technologies." In: Refusing, Limiting, Departing, In: CHI 2014 Workshop Considering Why We Should Study Technology Non-use, Toronto http://nonuse.jedbrubaker.com/wp-content/uploads/2014/03/NonUse_CHI_AlexLeavitt_final.pdf (last accessed: 03/07/2015).

Lenhart, A. (2010): Cell Phones and American Adults, Washington: Pew Research Center http://pewinternet.org/~/media//Files/Reports/2010/PIP_Adults_Cellphones_Report_2010.pdf (last accessed: 03/07/2015).

Ling, R. (2008): "Should we be Concerned that the Elderly don't Text?" In: The Information Society 24/5, pp. 334–341.

Loos, E./Haddon, L./Mante-Meijer, E.A. (eds.) (2012): Generational Use of New Media, Burlington: Ashgate.

Maxwell, J.A. (2005): Qualitative Research Design: An Interactive Approach, Thousand Oaks: Sage.

Neugarten, B.L. (1996): "The Young-Old and the Age-Irrelevant Society." In: Neugarten, B.L./Neugarten, D.A (eds.), The Meanings of Age: Selected Papers of Bernice L. Neugarten, Chicago: University of Chicago Press, pp. 47–55.

Östlund, B. (2004): "Social Science Research on Technology and the Elderly – Does it Exist?" In: Science & Technology Studies 17/2, pp. 44–62.

Oulasvirta, A./Wahlström, M./Ericsson, K.A. (2011): "What Does it Mean to be Good at Using a Mobile Device? An Investigation of Three Levels of Experience and Skill." In: International Journal of Human-Computer Studies 69/3, pp. 155–169.

Pinch, T.J./Bijker, W.E. (1987): "The Social Construction of Facts and Artifacts: Or How the Sociology of Science and the Sociology of Technology Might Benefit Each Other." In: Bijker, W.E./ Hughes, T.P./Pinch, T.J. (eds.), The Social Construction of Technological Systems: New Directions in the Sociology and History of Technology, Cambridge: MIT Press, pp. 17–50.

Salovaara, A./Lehmuskallio, A./Hedman, L./Valkonen, P./Näsänen, J. (2010): "Information Technologies and Transitions in the Lives of 55–65-Year-Olds: The Case of Colliding Life Interests." In: International Journal of Human-Computer Studies 68/11, pp. 803–821.

Sawchuk, K./Crow, B. (2011): "Into the Grey Zone: Seniors, Cell Phones and Milieus that Matter." In: WI: Journal of Mobile Media. http://wi.mobilities.ca/into-the-grey-zone-seniors-cell-phones-and-milieus-that-matter/ (last accessed: 03/07/2015).

Selwyn, N. (2003): "Apart from Technology: Understanding People's Non-Use of Information and Communication Technologies in Everyday Life." In: Technology in Society 25, pp. 99–116.

Tacchi, J.A./Slater, D./Hearn, G.N. (2003): Ethnographic Action Research: A User's Handbook, New Delhi: UNESCO.

Williamson, K./Asla, T. (2009): "Information Behavior of People in the Fourth Age: Implications for the Conceptualization of Information Literacy." In: Library & Information Science Research 31/2, pp. 76–83.

Wyatt, S. (2014): "Bringing Users and Non-Users into Being Across Methods and Disciplines." In: Refusing, Limiting, Departing, In: CHI 2014 Workshop Considering Why We Should Study Technology Non-use, Toronto. http://nonuse.jedbrubaker.com/wp-content/uploads/2014/03/Wyatt_Toronto_April_2014.pdf (last accessed 03/07/2015).

Wyatt, S./Thomas, G./Terranova, T. (2002): "They Came, They Surfed, They Went Back to the Beach: Conceptualizing Use and Non-Use of the Internet." In: Woolgar, S. (ed.), Virtual society? Technology, Cyberbole, Reality, Oxford, New York: Oxford University Press, pp. 23–40.

Yin, R.K. (2003): Case Study Research: Design and Methods, Thousand Oaks: Sage Publications.

Young, M./Foot, K./Gomez, R./Kinsley, R.P./Morrison, S./Weiss, P. (2014): "Towards a Research Agenda on Technology Pushback." In: Refusing, Limiting, Departing, In: CHI 2014 Workshop Considering Why We Should Study Technology Non-use, Toronto http://nonuse.jedbrubaker.com/wp-content/uploads/2014/03/pushback_workshop.pdf (last accessed 03/07/2015).

Older Women on the Game: Understanding Digital Game Perspectives from an Ageing Cohort

Hannah Marston and Sheri Graner-Ray

1. Introduction

There has been a substantial amount of activity in the game industry in relation to the deployment of digital games since their demise in the late 1950s and early 1960s. Conversely, within the early part of the 21st century, researchers have also taken an interest in digital games and their use within society for health, rehabilitation, societal engagement and playing for fun. The first computer game to be released by the game industry was *Spacewar* (1962), and from the beginning of the 1970s onwards, games were primarily accessible in public environments such as pubs and arcades (Simon 2009). This notion changed with the development of the Magnavox Odyssey (1972), which brought games into the home environment to facilitate a social relationship between the gamer and the games, but also amongst fellow players within the home. Moreover, the integration of computer games into the home is similar to the implementation of the television (Flynn 2003). While the television facilitated family togetherness and brought an entertainment medium, computer games also provided the family with a different form of entertainment, accessible in the home and not in public houses or arcades. Herman (2001) has provided an in-depth account of the videogame industry by highlighting the highs and lows experienced in the last forty years. Although this type of technology has been made available on the market for all audiences, it has primarily been accessed by tech-savvy individuals categorized as Generation X (1960s – early 1980s) and the Millennial Generation (1981-present) rather than by baby boomers, and by young males rather than by older women.

The concept of digital games for those in the childhood and youth period has primarily been male oriented (Laurel 2001), a tendency that changed in the 1980s as a result of several pioneers in the game industry: Brenda Braithwaite Romero[1], Roberta Williams[2] and Sheri Graner Ray[3]. The array of practical and theoretical work undertaken in this field by designers, critics and scholars has recognized that many games do not appeal to the female audience. Solutions have been sought to overcome this issue to enhance a greater understanding of the needs and requirements of female gamers in relation to content and access.

The use of digital games by 'other' audiences has been exponential (Bleakley et al. 2013; Hall et al. 2012; Marston/Smith 2012; Miller et al 2013) with a strong focus of facilitating intergenerational game play via devices such as the Nintendo DS handheld, the Wii, and the Microsoft Kinect consoles. The average age of female US gamers has changed considerably over the last 15 years based on the statistics published by the ESA. The average age of a female gamer in 2004 was 29 years (ESA 2004), and eleven years later, the average age of a female gamer is now 43 years, whereas the average age of a male gamer is 35 years old (ESA, 2015).

Since 2004 there has been a growing proportion of female gamers (ESA, 2004); thus illustrating a growing market which is now perceived to be an entertainment medium for all people and ages in society, not just for men or boys. Little work has provided an insight into older person's preferences of digital games (Marston 2013a; 2012); even less is known about digital game preferences of older women.

---

1 | Brenda Brathwaite Romero has worked in the games industry for 34 years and has credits on 22 game titles. Ms Brathwatie Romero is best known for working on the Wizardy role-playing video games.

2 | Roberta Williams is a game designer, writer and a co-founder of Sierra On-Line (later known as Sierra Entertainment). Ms Williams pioneering work in the games industry stems from the graphic adventure games which include game titles such as Mystery House, the King's Ques and Phantasmagoria.

3 | Sheri Graner Ray has worked for a series of video games companies including Electronic Arts, Origin Systems, Sony Online Entertainment and Cartoon Network. Ms Graner Ray has worked on game titles such as Star Wars Galaxies, Ultima and Nancy Drew. In 2005 she received the International Game Developers Association Award for Community Contribution for her work on women's interests in the games industry. At present she currently serves as the Executive Char for Women in Games International.

However, with industry developments such as the development of the Wii or the Kinect console, it is unlikely that digital games will stop being utilized by older people based on the advantages highlighted by previous studies (de Schutter 2010; de Schutter/Vanden Abeele 2010; Marston 2013a, 2010; Voida/Carpendale/Greenberg, 2010; Voida/Greenberg 2009). The authors suggest these consoles are popular with older adults because of the type of interaction utilized by the individual consoles – motion and gesture/ speech recognition. This makes it easier for older adults to engage with the game environment than playing on the Sony Play Station console, which uses the traditional game pad for game play interaction which older adults may find difficult to use due to age related conditions such as arthritis or unfamiliarity with the controller (Marston 2013b).

Figure 1:Displays the Increase of Gamers (by Age) since 2004

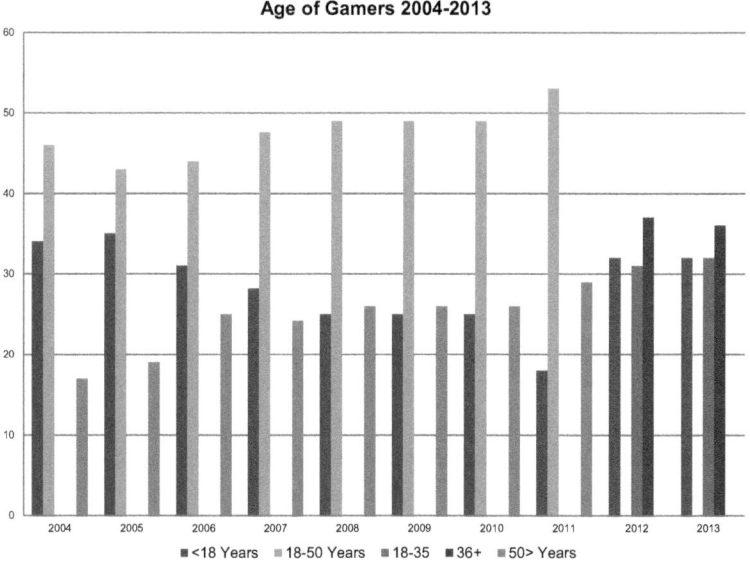

It is anticipated that the ageing population will reach unprecedented numbers in the next fifty years (European Commission, 2007). Based upon the estimated statistics and in conjunction with the digital gaming developments, it is hypothesized that there is the possibility that game technologies could facilitate independent living, user engagement, well-being and rehabilitation to improve one's quality of life.

Within the fields of gender and game studies, older women have tended to be ignored (Krekula 2007). To the authors' knowledge, there is no work that focuses specifically on digital game preferences of older women. Besides, little research so far examines digital games from a gender and gerontological perspective. Moreover, there have been no published data referring to the number of hours spent by older women and their game playing habits in conjunction with the type of genre(s) they prefer. Identifying and understanding the preferred type of content of older women may aid researchers and professionals with the knowledge for future digital game design and developments. The work discussed in this chapter attempts to answer the following question: what are the gaming preferences of older women?

2. A Gender-Based Perspective on Gaming

Recognition of the importance to examine preferences and engagement of digital games by older adults has begun; yet, to date, its focus has been limited. Current research has examined the utilization of digital games to facilitate quality of life (Goldstein et al. 1997), the effects of cognition and reaction time while interacting with digital games (Basak et al. 2008), and design interaction, intergenerational gaming and requirements of older adults for prospective game design (de Schutter 2010; de Schutter/vanden Abeele 2010; IJsselsteijn et al. 2007; Marston 2013a; Marston/Smith 2012; Pearce 2008; Voida/Carpendale/Greenberg 2010; Voida/Greenberg 2009). Researchers have examined the relationship between gender and digital games primarily focusing on younger audiences (Carr 2006; Cassel/Jenkins 1998 Heeter et al. 2004; Jenkins 1998; Kafai 1996; Krotoski 2004; Pratchett 2005; Taylor 2003, 2006).

Game concepts for younger audiences designed by males often integrate elements of violence, which in turn became less appealing for females (Kafai 1996). Several studies have identified game elements which should be considered during the development process of games to target a female audience. The elements listed below are deemed important to female players and digital games, which include the following (Carr 2006; Cassel/Jenkins 1998; Graner Ray 2004; Heeter et al. 2004; Jenkins 1998; Krotoski 2004).

- Character development;
- Collaborative play;
- Narrative;
- Puzzle;
- Exploration;
- Role playing;
- Game experience to reflect real world experiences;
- Social interaction;
- Indirect competition;
- Team competition;
- Cooperative play; and
- Flexibility to choose and make decisions.

Traditionally, female character development and placement in games has taken on submissive roles such as the damsel in distress, or a princess in need to be rescued by a male character. One of the most internationally well recognized examples can be found in the *Super Mario Bros* (1985) game. The gamer may select either one or two male characters and must complete a series of levels to secure the release of the princess; supporting previous research that illustrates representation is often male dominated (Laurel 2001). Therefore, there is an opportunity within game design to encompass a strong female protagonist: seizing this opportunity, to broaden the appeal of the game to a female audience in addition to widening the scope of game ideas to welcome older women, Bryce & Rutter among others note that:

"[…] female game characters are routinely represented in a narrowly stereotypical manner; for example, as princesses or wise old women in fantasy games, as objects waiting on male rescue, or as fetishised subjects of the male gaze in first person shooters" (2002: 246).

Taylor (2003) explored how the female representation of avatars can affect female gamers, highlighting that a common feature within games can originate from male and female avatars and the hyper sexualisation of the characters. For example, female avatars are specifically sexual, and while "chest and biceps on male characters act as symbolic sexual characteristics, they are simultaneously able to represent power… large breasts only act as sexual markers" (41).

Graner Ray (2004) critiques / argues against the perception of 'heroes' who are identified through physical traits of "unrealistically large breasts situated highly on the chest, a waspish waist, and a prominent, well-rounded derriere" (102), which positions the character to be "[...] young, fertile and ready for sex" (102). This notion is perceived acceptable by players who have the desire to be a hero. For example *Xena, Warrior Princess* (Renaissance Pictures) is shown to be a strong and sexy character and not hyper sexualized in the television program bearing the same title. Her physical body proportions were not exaggerated and yet represented a heroic character. Furthermore, Graner Ray notes how female gamers reported an increase in confidence through the representation of an avatar (2004: 105).

The evolution of Lara Croft (1996-2013) has changed considerably. Designers/developers have continued to hyper-sexualise the character: for example, her stomach and breasts are uncovered in the latter character developments, more so than in the earlier images. During game play, the camera angles highlight the hyper-sexualized character development of Lara Croft, to emphasize her bust. As Graner Ray states, "Even when on her back and covered in the sand, her very large bust line stood straight up like two large missiles" (2004: 33). With this in mind, Lara Croft may be perceived as a negative role model. In the 1990s, characters such as Buffy the Vampire Slayer and Lara Croft rose to popularity amongst the 'laddette' culture whereby women would play the 'lads' at their own game (Kennedy 2002). With this in mind, Lara Croft's over exaggerated body may be perceived as a male fantasy.

In addition to the limited work (Krotoski 2004; Pratchett et al. 2005; Taylor 2003, 2006) focusing on gender representation, there is still little information focusing on digital device ownership, behaviour and preferences by older adults – in particular older women – with the exception of Pratchett et al. (2005), who reported the use and ownership of digital devices, digital game consoles and user preferences across age from UK users. The survey highlighted that 40 per cent of the adults aged 51-65 years owned games consoles and 18% owned a handheld game device (Pratchett et al. 2005).

Pratchett et al. (2005) notes how the digital game preferences display similarities and differences between the sexes. For example, simulations[4] and massively multiplayer online[5] (MMOs) games such as *World of Warcraft* (WOW,

---

4 | Simulation games aim at simulating real life activities – for example, in a business or training situation. An example of a simulation game is SimCity (Maxis).

5 | Massively multiplayer online games can facilitate a large number of players within one environment, simultaneously engaging with one another or executing a variety of different tasks.

Blizzard Entertainment) were equal for both genders, unlike strategy[6] and role playing games[7] (RPGs), which were slightly surpassed by males, while women preferred to play music/dance, puzzles/ quizzes, and classic board/ games.

Winn & Heeter (2009) conducted a study of game playing habits among college students and identified that male gamers played games for a minimum of 60 minutes or more per day. In contrast, female gamers reported to have less time, and preferred to play for shorter time periods. This results in a preference of different game genres being played – such as casual instead of first-person shooter (FPS) – according to the time needed to play FPS games, as opposed to puzzle games, which a player can pick-up and play over a 30-minute period, whereas playing an FPS game requires more time. Moreover, "women have smaller chunks of free time than males" (Winn & Heeter 2009: 4), assuming that women have to conduct additional day-to-day activities; resulting in the female students preferring to play casual games[8] rather than games which take larger chunks of time to play.

3. GERONTOLOGICAL PERSPECTIVES ON AGEING

The following section provides different gerontology theories that have served the field to build and enhance understanding of the life course with regard to changing the behaviour of older people throughout the lifespan. Havighurst's Activity Theory (1968) as seen in "Disengagement and Patterns of Aging" focuses on the individual's adjustment and adaptation, whereby a positive relationship is formed between an activity and life satisfaction. This is similar to the expression 'use it or lose it', which has been linked to the Brain Training games aimed at delaying or mitigating the effects of dementia (Basak et al. 2008; Clark/Lanphear/Riddick 1987).

Kart & Manard (1981) suggest there is a positive relationship between activity theory and life satisfaction, whereby the decrease of social loss results in a decrease in life satisfaction. They contend that the" most successful

---

6 | Strategy games such as Command & Conquer facilitates the player(s) decision making to affect the overall outcome.

7 | RPGs allow the player to take on a specific role within a game such as Doom (EA Games).

8 | Casual games are "A game intended for people for whom gaming is not a primary area of interest" (Sheffield 2008).

aging (adjustment) occurs for those persons who stay active and resist the consequences of changes that equate with losses" (449). Lemon/Bengtson/Peterson (1972, 1981) reported criticism of this theory and identified that it was perceived as being "a simple, linear model for predicting life satisfaction" (1972: 520). Lemon/Bengtson/Peterson 1972, 1981) identified this theory to be inadequate to portray the intricacy of interaction between individuals and social situations (Howe 1987).

In 1961, Cumming & Henry posited disengagement theory as a process whereby an individual's age is a mutual agreement between the person and society that commences via the process of withdrawal from social roles and relationships with society. It is suggested "that it was beneficial for both the aging individual and the society that such a disengagement takes place in order to minimize the social disruption caused at the older person's eventual death" (Morgen/Kunkel 1998: 274). Although this theory focuses on the eventual departure from earth, it also corresponds to individuals transferring from the work place into retirement. There is criticism of disengagement theory: for example, Kastenbaum (1993) proposed that the ageing process should be positive and entail an engaged lifestyle.

However, Morgen & Kunkel (1998: 275) noted that the aforementioned theory was not necessarily fixed and this was the ethos of disengagement theory; it soon became a normative statement ("People *should* disengage") rather than as a description of reality ("As they age, people *do* disengage"). Hochschild (1975) contends that the notion of disengagement is conventional and foreseeable, as the timing and nature varies amongst an ageing population. Paoletti (1998) shows how older adults construct age identity through social construction. The participants perceived themselves "as active and effective older adults" (Lin/Hummert/Harwood 2004: 263), in comparison to other older adults who maybe" lonely, sick, and dependent" (Lin/Hummert/Harwood 2004: 263). The adults in the study conducted by Paoletti (1998) distinguished themselves and formed a perspective of a" positive group identity by verbally distancing themselves from other older people" (Lin/Hummert/Harwood 2004: 263).

The concept of the life course as structured towards growth and maturing is reinforced and emphasised: the stages within the life course are childhood, adolescence, adulthood, middle age and old age. Within these categories, there are norms which society perceives to be acceptable. For example, a woman aged between 20-40 years giving birth would be socially acceptable, while a woman aged 50+ having her first child would not be perceived to follow the norm with respect to the biological processes. Taking into account the life course, the implementation of digital games was initially perceived as a medium for boys (Laurel 2001) as well as men who perceived it as a form of leisure (Bryce/Rutter 2003; Cassell/Jenkins 1998), a view that is similar to that of Haddon (1992) and Turkle (1984). However, it is hypothesized with the anticipation of ageing populations, the implementation of digital games and technology into one's life at the later stages of the life course has the potential impact to transform and facilitate the user's life in many ways, such as in health and well-being, learning new skills, communication with friends and family, and information seeking.

Feminist gerontology is built upon familiar aspirations comprising of the "development of social consciousness about inequalities, utilization of theories and methods that accurately depict life experiences, and promotion of change in conditions that negatively impact older people or women" (Reinharz 1986: 504). Garner (1999) outlines how feminism and gerontology have similar foci: recognising women and older adults as individuals receiving equal treatment. The work conducted within gerontology in recent years has primarily focused on older adults as a whole without making distinctions of sex (Garner 1999): in particular, the primary study of men as subjects was criticized within the medical field of gerontology (Garner 1999). Garner (1999) suggests that the increase of feminist gerontology could be due to the ageing process of feminists themselves. Reviewing the literature surrounding digital games from a gerontological perspective, there is little or no published work that makes the distinction between older women and men. The relationship between feminism and gerontology continues and, as Garner (1999) notes, feminism and gerontology attempt to build upon mechanisms for both social change and individual empowerment.

This section has provided an overview of different gerontological theories which display how adults in later life and in conjunction with feminist perspectives can be portrayed and associated to work in the 21st Century in particular where technology may have an influence within one's life. So far, empirical research is missing in the fields of gender studies and gerontechnology with a primary focus of older women's digital game preferences. In the next section, the authors outline a study which formed a sub-domain of Marston PhD work, in addition to the results, discussion and conclusion.

4. EMPIRICAL DESIGN

The data presented in this paper were collected in 2005 as part of Marston's PhD thesis (2010). A total of 28 participants were contacted and recruited at Teesside University and the Psychology and Communication Technology (PaCT) Lab at Northumbria University. Twenty-four participants (male and female) aged between 53-75 years were recruited for the study. The participants were provided with an informed consent and the resulting data are reported in the following sections.

The purpose of this workshop conducted by Marston was to create individual digital game concepts through a paper and pencil exercise. Marston (2012) provides an in depth detail and break down of the game design workshops that enabled the participants to create their own game concepts through a step-by-step approach. Workshops were conducted and included: (a) an introduction to the workshop and the purpose of data collection, (b) getting started – which focused on the participants prospective game idea, (c) how to play – which focused on how prospective players would interact with the game (e.g., keyboard, gesture, mouse) and, finally, (d) the finale – which covered the costing of the game, additional target audiences, and marketing of the game. The participants completed the survey once the workshop had finished.

A quantitative survey was developed and included several items: console ownership, whether the participant(s) would consider game playing, preferred game genre(s), frequency and length of game playing, purchasing habits, method of learning, and demographics.

5. RESULTS

Quantitative Results

The personal computer (PC) was primarily owned by female participants and a small percentage of the participants reported to owning a digital game console. A variety of hobbies were reported by the female participants, which primarily included walking, doing arts & crafts, playing cards, dancing, watching television, going to the cinema, and playing board games. A variety of preferred game genres were reported by the female participants, which included games in the puzzle, strategy, adventure, shooter and platform genres. Participants primarily reported the type of game genres they would consider playing, which included: puzzle strategy, board games, sport, and adventure. Participants primarily reported to teach themselves how to play games, in addition to learning via their grand/children and in educational classes. A positive trend was shown by the female participants willing to play games and who were also positive in playing games relating to a hobby or interest.

Qualitative Results

During the workshops the participants were asked to record a series of verbs they would incorporate into their game concept: different types of verbs recorded by the participants, initially forming a basis for their game concepts. Throughout digital games, one will always find at least one or several imbalances such as human or alien, good or bad; several imbalances were recorded by the participants, which they would want to implement into their game concepts. The pleasure that can be experienced by gamers varies during their digital game engagement and can include achievement, gratification, and satisfaction of completing a level, and attaining a specific object needed to complete a task. However, there were variations depending on the type of game – such as massively multiplayer online role-playing game (MMORPG), which facilitates conversation and friendship amongst players; whereas other rewards for completing a task or reaching the next level. For some, mastering a particular skill is more important and this may be attained through team work or playing as a single player.

Identifying the Big Picture

Taking into account the different keywords and sections previously discussed several findings associated to game ideas can be discovered (c.f Table 1). Taking this a step further and utilizing the metacritic[9] website, which follows a similar undertaking by the games industry, the game ideas were identified via discourse analysis by highlighting themes primarily based on keywords such as education or action (c.f. Table 2).

---

9 | http://www.metacritic.com/games.

Table 1: Displays Several Game Concepts highlighted by the Participants

Big Picture (concepts)
• Garden design, variety of styles
• Travel the world, offer educational information, sites of interest
• A quest, problem solving, end up with a final cross stitch based upon
• Puzzle, set in a work environment, able to distress, promote colleagues
• Human body, a blood cell travels around the body, and learn about different sections
• Tour around a country/county, visiting historical sites
• A situation in history with fantasy. Variety of different peoples (invaders, towns folk etc). Have to collect supplies
• Nature trail, identify animals, flowers bird songs
• Social situation, play different roles within society
• Sport – track and field events
• Travel to historical place(s) learn more about a place
• Escaping from a movie which has obstacles, similar to a maze
• Travel around the world, a period of history and offers different questions etc to gain entry/travel throughout • Museum work/meeting friends

Table 2: Game Genre categorized from the Concepts created by Female Participants

Game Genres from Concepts	
Genre	Frequency[1]
Educational	8
Puzzle	2
Travel	5
Adventure/Quest	1
Strategy	1
Fantasy	1
Role responsibility	1
Sport	1
Other	5

Keywords were grouped together (c.f. Table 3) to identify a common thread, which enabled the identification of themes. Keywords were highlighted via discourse analysis and categorized according to the section(s) that were presented to the participants.

Table 3: Displays the Themes identified from the Keywords collected during the Workshops

Verbs		Theme	Pleasure		Themes
Reading Playing Modelling Working Flying Counting Measuring	Walking Eating Listening Drinking Observing Sewing Attaining	Recreation	Rule breaking Annoying people Expressing feelings Calm/relaxed Friendship	Amusement Frustration Exhilaration Fascination Relief Challenge	Mentality
Swimming Riding Climbing Skiing	Sailing Shooting Boxing Flying	Play	Learning Challenge Engagement	Mastery of skill Gratification	Pedagogy
Fighting Shooting Attainment Building Modelling	Missing class Labelling Fighting	Regulation	Menus/touch Screen Buttons Fun	Calm/relaxed Expressing feelings Friendship	Interaction
Learning Reading Speaking Researching	Planning/ designing Attaining	Pedagogy	Experience without being there Work colleagues Friendship Beauty		Environment
Looking Smiling	Shouting Observing	Sentiment	Achievement Fun Gratification	Satisfaction Regret of outcome Winning	Atonement
Labelling Measuring Keeping mind active Observing	Learning Hiding Looking Walking Researching	Exploration			
Imbalances		Theme	Game Space		Themes
Order/chaos Good/bad Relaxed/tired Un/prepared Young/old	Uncaring Bogus clues Un/common Light/dark Nice/nasty	Human intervention	Realistic 1st person view 3rd person view Cycles of seasons		Perspective
Male/female Human/other People (good/bad) Un/caring Un/prepared	Less gore, more learning Un/interesting	Human personality	Sound effects On screen tools (compass) Pictures & words		Representation
Preserved/natural Human/alien Un/edible Short/long distance Smooth/rough distance Transport – lack of connection Transport – to get from A to B	Weather Foliage, seasons, football pitch/ formality/ adventure playground	Physical Space Intervention	Ease of movement in environment, clues provided when hovering over information		
What is the Point?			Theme		
Improve society Have fun – the 'what if' situation Act out hidden emotions			Role play		
Pass the time Strategy Education Achievement			Mentality		
To reach a certain point within a specific time Win/lose Achievement Satisfaction			Atonement		

6. DISCUSSION

This chapter has aimed to explore digital gaming preferences by older women and the results provided by the participants have shown positive perspectives for this cohort, which is underrepresented in the literature of game studies and gerontology.

The participants reported a game idea that has an educational focus primarily based around travel and/or historical facts/events. The concept of a game associated to travelling may have been because some participants were close to or had reached retirement or for some, or because restrictions were imposed due to health and/or financial reasons. Having a game that encompasses a travel element could help alleviate these issues. This could be associated to activity theory, whereby a person who has more time and less financial restrictions has the ability to conduct different activities such as travelling. Thus, the game would provide further/additional life satisfaction to one's life through travel and seeing places of interest. The data implies the participants still wish to maintain a productive activity that may include self-improvement.

The game ideas do not necessarily fit within the confines of the games industry classification, and the female participants reported positively to playing a game relating to their hobby or interest. The data suggest that the concepts may provide a new area for game development, and that being aware of the perceptions and requirements of older women may broaden their participation and interest.

Primarily, female participants owned a PC, which echoes similar findings to Pratchett et al. (2005), who identified that 84 per cent of older adults owned a PC. Equally, Orlov (2011) reports PC ownership by adults aged 74+ years was 41 per cent. Game console ownership by the female participants was not so common and the type of consoles owned by the participants included a PS1, Nintendo Game Cube and Nintendo Game Boy handheld.

The primary approach to learning how to play games varied and the female participants reported teaching themselves and being taught by a grandchild. Attempting to understand how older adults prefer or have learnt how to play games could be vital when introducing the concept for the first time; this is of particular concern when games are utilized within a healthcare or long-term care setting. The length of game playing varied and the majority of the female participants reported to have been gaming for more than one year. Female participants primarily played games on a weekly and daily basis. Similarly, the results presented by Pratchett et al. (2005) reported that 30 per cent of female respondents played games once to twice a week, 23 per cent played once to three times a month and 31 per cent played less often.

The preferred game genre played by female participants was the puzzle genre. Similarly, this is supported by Pratchett et al. (2005) who also reported that female participants in the respective study preferred playing puzzle games in addition to playing strategy games. Additional genres were also reported by the female participants and included adventure, platform, shooter and sports games.

Several hobbies and interests were identified by the female participants, in particular arts and crafts and walking, followed by card games, dancing, cycling, needlecraft and gardening. The game concepts were designed accordingly to the participant's hobbies and interests (c.f. Table 2). The participants showed a positive response to playing a game related to their hobby or interest, although participants did report that they were unsure of wanting to play a game associated to their hobbies or interests.

A variety of pleasures were reported by the participants, including achievement, fun, gratification, satisfaction, learning, engagement, challenge and frustration. These results imply that the participants would like to experience a variety of pleasures during their gaming experience. The preferred mode of communication by the participants is traditional: for example, the female participants would like to utilize a keyboard, a joystick, a mouse or a touch screen to engage with their game. Due to the time of the study being conducted, gesture or motion control was innovative or not available, whereas now this is quite a common approach to game engagement.

These results should take into account that the participants did report themselves as gamers and that they may have pre-conceived approaches to game engagement. At the time of this study, neither the Wii nor the Kinect consoles were available on the market. The authors consider that if this study had been conducted after the release of this hardware, the method of communication (interaction) could have been different. The type of visualization reported could include simplicity, images and words, easy movement and clues to the user while hovering over the information tag. Participants reported that the primary purpose of playing games is to pass the time, to experience a 'what if' situation, and to act out hidden emotions. We infer that the participants would like to experience similar emotions to those often experienced by younger audiences.

Figure 2: Displays the Overarching Relationship between the Themes and Sub-themes

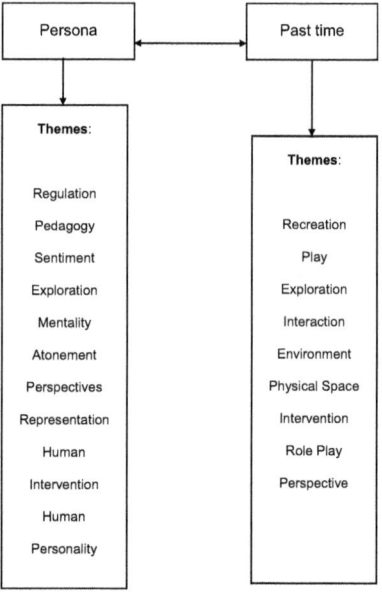

Based on the data analysed, two overarching themes, *persona* (e.g. human feelings, emotions, cognitive, and physical attributes) and *pastime* (e.g. human action(s), conducted in different physical spaces, with/out partnerships, and differing evaluations and assessments), have been established from the data collection/analysis: these have been correlated into two overarching themes to illustrate potential relationships (Fig 2), in particular to specific game requirements which can provide a primary sense of role play, pedagogy, and/or play/recreation. The participants seem to prefer games connected to several objectives and include personality, pedagogy, perspective, mentality, sentiment, human intervention (rehabilitation/health/well-being), and personality. Although the game concepts were individually designed, we believe that the connections between the overarching themes and sub-theme/keywords denote similarities between the participants' concepts once they have been broken down.

The section 'what is the point' highlighted role-play as an important factor for the female participants. Role-play consists in a person undertaking a different personality other than his/her own: for example, a person may work in a factory but, when playing digital games, s/he may choose to play the role of a god/dess. We propose that this theme could be correlated to older women's roles and experiences within society, but transferred into the digital environment as they age. For example, the underpinning notion of disengagement theory is for one to withdraw from society as one age; however, undertaking a particular role within a digital game could provide a different experience for the participant undertaking a role in the real world.

The concepts derived from the verbs and imbalance section can also relate to activity theory: for example, the utilization of an activity can prove to have a positive effect on a relationship and life satisfaction. The results suggest that identifying the keywords situated as sub-themes – such as atonement, mentality, sentiment, exploration, play and recreation – has the potential to enhance positivity between digital games and life satisfaction. Therefore, older women who have retired from employment can engage with regular activities they have enjoyed throughout their life course but who are required to undertake rehabilitation due to ill health, an activity that may bring a different level of satisfaction in relation to such activities.

The life course theory (Neugarten/Hagestad 1976) provides an alternative definition to ageing and constructs the concept of ageing from biological, social, psychological and cultural perspectives, rather than focusing on one facet. The adoption of games in conjunction with the life course theory has an all rounded perspective that does not focus on an individual facet of the ageing process, such as the image of an elderly person who lives in a residential home and who has little contact with residents. Playing a game such as *Wii Sports Bowling* may facilitate an increase in social interaction, as shown by Harley et al. (2010). A person who attends a local social event at the church hall or community centre may choose to attend more frequently having the knowledge that other attendants are also members of the same bowling team. Older women who choose to interact with this type of technology may bridge the gap (if it is the case) with grandchildren, who are aware and use digital technology for recreational purposes. While our ageing populations may be aged 65+ years old, socially and psychologically they may feel younger and perceive themselves to be forty or younger.

7. IMPLICATIONS FOR THE FUTURE

Earlier work (Marston 2013a; Voida/Carpendale/Greenberg, 2010; Voida/Greenberg 2009) has formed the initial basis for continual development of games to be utilized for an ageing population. This area of research is crucial for both academia and industry, since digital games have the potential to provide users with positive benefits. Based on the current ageing population, individuals are seeking alternative approaches to assist older adults and elderly citizens in their daily lives and activities. Taking into account the themes highlighted, one should consider the implementation of different facets of digital games for older female users who wish to explore, learn or undertake a different responsibility to those they have previously undertaken.

The statistics from the ESA (2004; 2014) have highlighted the growth of older persons playing games and the type of game genres played, but in more recent years these data have not been recorded. Yet, it is known that the ageing population will continue to grow (European Commission 2007) and having this information is important to understand who is playing what type of game genres in order to assist the academic and industry communities in providing suitable, beneficial, engaging and funny entertainment. This chapter has demonstrated that it is possible for female users to interact with a game to improve their knowledge about different countries, cultures and languages, something that may not have been possible before due to other commitments. Additionally, taking on a role which is not the norm or which is not accessible within society and being able to change their personality to a character within the game environment may be attractive to female users.

Utilizing digital games in community centres, or residential homes has the potential to provide a different form of socialisation amongst friends and peers. Engaging with games which have suitable and purposeful content for older adults may prove fruitful in the future and prevent persons from displaying signs of disengagement theory, as shown in the work of Paoletti (1998) and Lin, Hummert & Hardwood (2004). Although one may consider the loss of face-to-face engagement whilst playing digital games, there is evidence of bowling leagues being created within senior and residential environments (Harley et al. 2010; Jung et al. 2009). The creation of such events has limited social interaction but has enhanced socialization, and it is suggested this concept could be transferred across many communities – provided there is education, communication and demonstration of technology by community volunteers, family and peers.

The concepts presented in this chapter have demonstrated how the female participants require a variety of games that combine educational elements in addition to a preconceived purpose. One genre which has the potential to provide an ageing society with an alternative option is the genre 'other'. During data analysis, some game ideas were not easily categorized and were therefore placed into the genre 'other'. Elaborating on this notion by consulting with older women during the initial design phases and through further workshops and one-to-one interviews could produce a series of suitable ideas.

The main limitations in this study are its small sample size, resulting in the lack of multivariate analysis conducted between groups or gender variations to identify any significance. One-to-one interviews were not conducted and this is something that should be taken into consideration for future explorations.

Further work in the form of focus groups and participatory design would be required to fully identify and understand the essential needs and requirements found in this initial work. Implementing a greater in-depth qualitative approach has the potential to build upon the initial concept of the participant to identify their reasoning and choices for creating the concept. In turn, prototypes could be developed and evaluated, enabling user engagement and providing feedback to ascertain whether the key elements reach the goals and needs of the users.

Technology is still perceived as a masculine entity and the concept of engaging in digital games is not different. There has been a substantial growth of games targeted at girls and a greater focus on the preferences that are important for this cohort, yet there is little understanding of games for older women. We propose that future studies should consider data sets based on gender needs, preferences, ownership and usage, thus enabling a mapping approach of technology and gaming for future development and building up a form of database to assist in the future ageing demography.

8. Conclusion

To sum up, the focus of this chapter was based on the design of digital game concepts by older women, with a view to broadening and providing a greater understanding of game preferences. Although there have been advancements in this area, little work has focused on older women. Taking into account the data set and the discourse analysis approach, several themes have been highlighted and the authors have attempted to display the connections based upon the preferred content preferences, resulting into two overarching themes (persona and past time). With this in mind, the data suggest that the participants would like to experience a variety of emotions, gratifications and roles that may not have been available to them in earlier years of their life course. Although one could say this is not dissimilar from gamers themselves, who play games so they can be a rock star or a dragon, from a female and life course perspective it refers to the fact that they may want to play roles that were not available to them due to preconceived notions of society or upbringing. Nonetheless, with the purpose of using digital games for social engagement, rehabilitation and the potential to bridge intergenerational relationships, the data presented could serve as an initial step to fully comprehend the needs and requirements of older female users.

Acknowledgements

I would like to thank the participants for contributing with their time to take part in this study.

References

Basak, C./Boot, W.R./Voss, M.W./Kramer, A.F. (2008): "Can Training in a Real-Time Strategy Video Game Attenuate Cognitive Decline in Older Adults?" In: Psychology and Aging 23/4, pp. 765-777.

Bleakley, C.M./Charles, D./Porter-Armstrong, A./McNeill, M.D.J. /McDonough, S.M./McCormack, B. (2013 Epub): "Gaming for Health: A Systematic Review of the Physical and Cognitive Effects of Interactive Computer Games in Older Adults." In: Journal of Applied Gerontology, 34/3, pp. 166-89 Doi:10.1177/0733464812470747

Bryce, J./Rutter, J. (2003): "Gender dynamics and the social and spatial organization of computer gaming." In: Leisure Studies 22, pp. 1-15.

Carr, D. (2006): "Games and Gender." In: Carr, D./Buckingham, D./Burn, A./Schott, G. (eds.), Computer Games, Text, Narrative and Play, pp. 162-178. Cambridge, Policy Press.

Cassell, J./Jenkins, H. (1998): "Chess for Girls? Feminism and Computer Games", In: Cassell, J./Jenkins, H. (eds.), Barbie to Mortal Kombat, Boston, Massachusetts: The MIT Press, pp. 2-45.

Clark, J.E./Lanphear, A.K./Riddick, C.C. (1987): "The Effects of Videogame Playing on the Response Selection Processing of Elderly Adults." In: Journal of Gerontology 42/1, pp. 82-83.

Cumming, E./Henry, W.E. (1961): Growing old: The process of disengagement, New York: Basic Books.

de Schutter, B./Vanden Abeele, V. (2008): Meaningful play in elderly life. In: 58th Annual Conference of the International Communication Association "Communicating for Social Impact" (conference paper), Montreal, Quebec, Canada, Le Centre Sheraton, May 22-26.

de Schutter, B. (2010): "Never Too Old to Play: The Appeal of Digital Games to an Older Audience." In: Games & Culture 6/2, pp.155-170.

ESA. Entertainment Software Association. (2004): "Essential Facts about the Computer and Video Game Industry", 2004, http://www.theesa.com/files/EFrochure.pdf.

ESA. Entertainment Software Association. (2014): "Essential Facts about the Computer and Video Game Industry" 2014, http://www.theesa.com/wp-content/uploads/2014/10/ESA_EF_2014.pdf

ESA. Entertainment Software Association. (2015): "Essential Facts about the Computer and Video Game Industry" 2015, http://www.theesa.com/wp-content/uploads/2015/04/ESA-Essential-Facts-2015.pdf

European Commission. (2007). "i2010: Independent Living for the Ageing Society", Retrieved from http://ec.europa.eu/information_society/activities/ict_psp/documents/independent_living.pdf, accessed June 2015.

Flynn, B. (2003): "Geography of the Digital Hearth. Information." In: Communication & Society 6/4, pp. 551-576.

Garner, D.J. (1999): "Feminism and Feminist Gerontology." In: Journal of Women & Aging 11/2-3, pp. 3-12.

Graner Ray, S. (2004): "Gender Inclusive Game Design: Expanding the Market." Charles River Media.

Goldstein, J./Cajko, L./Oosterbroek, M./Michielsen, M./van Houten, O./Salverda, F. (1997): "Videogames and the elderly." In Social Behavior and Personality 25/4, pp. 345-352.

Haddon, L. (1992): "Explaining ICT Consumption: The Case of the Home Computer," In: Silverstone, R./Hirsch, E. (eds.), Consuming Technologies: Media and Information in Domestic Spaces, London: Routledge, pp. 82-96.

Hall, A.K./Chavarria, E./Maneeratana, V./Chaney, B.H./Bernhardt, J.M. (2012): "Health Benefits of Digital Videogames for Older Adults: A Systematic Review of the Literature." In: Games for Health Journal 1/6, pp. 402-410.

Harley, D./Fitzpatrick, G./Axelrod, L./White, G./McAllister, G. (2010): "Making the Wii at home: Game play by older people in sheltered housing." In: Lecture Notes in Computer Science 6389, pp. 156–176.

Havighurst, R.J./Neugarten, B.L./Tobin, S.S. (1968): "Disengagement and patterns of aging." In: Neugarten, B.L (ed.), Middle age and aging: A reader in social psychology, Chicago: University of Chicago Press, pp. 161-172.

Heeter, C./Chunhui, K./Egidio, R./Mishra, P./Graves-Wolf, L. (2004): "Do Girls Prefer Games Designed by Girls?" Comm Tech Lab, Michigan State University, Retrieved from http://spacepioneers.msu.edu/girls_as_designers_spring_survey.pdf, accessed June 2015.

Herman, L. (2001): "Phoenix: The fall and rise of Videogames" 3rd Edition. Rolenta Press, Springfield, NJ, USA.

Hochschild, A.R. (1975): "Disengagement Theory: A Critique and Proposal." In: American Sociological Review 14/5, pp. 553-569.

Howe, C.Z. (1987). Selected Social Gerontology Theories and Older Adults Leisure Involvement: A Review of the Literature", Journal of Applied Gerontology, 6, pp.448-463, doi: 10.1177/073346488700600407.

IJsselsteijn, W./Nap, H.H./de Kort, Y./Poels, K. (2007): Digital game design for elderly users. Proceedings of the 2007 conference on Future Play. Toronto, pp. 17-22.

Jenkins, H. (1998): "Complete freedom of movement: Video Games as Gendered Play Spaces." In: Salen, K./Zimmerman, E. (eds.), The Game Design Reader, A rules of Play Anthology, pp. 330-363. Cambridge, MA: The MIT Press.

Jung, Y./Koay, J.L./Ng, S.J./Wong, L.C.G./Lee, K.M. (2009): "Games for a Better Life: Effects of Playing Wii Games on the Well-Being of Seniors in a Long-Term Care Facility." In: IE '09 Proceedings of the Sixth Australasian Conference on Interactive Entertainment. ACM, New York, pp. 0-5.

Kafai, Y.B. (1996): "Gender Difference in Children's Construction video Games." In:. Greenfield, P.M./Cocking, R.R. (eds.), Interacting with Video, Norwood, NJ: Ablex Publishing, pp. 39-66. Available at https://www.gse.upenn.edu/c4ls/sites/gse.upenn.edu.c4ls/files/pdfs/kafai.pdf, accessed [July 2015]

Kart, C.S./Manard, B.B. (eds.) (1981): "Aging in America: Readings in social gerontology" (3rd ed.). Palo Alto, CA: Mayfield. http://www.sagepub.com/moody6study/study/articles/controversy1/Howe.pdf, accessed June 2015.

Kastenbaum, R. (1993): "Disengagement Theory." In: Encyclopaedia of Adult Development. In: Robert Kastenbaum, (ed.). pp. 126-130. Phoenix: Oryx Press.

Kennedy, H.W. (2002): "Lara Croft: Feminist Icon or Cyberbimbo? On the Limits of Textual Analysis." In: Game Studies: International Journal of Computer Games Research. Retrieved from http://gamestudies.org/0202/kennedy/, accessed June 2015.

Krekula, C. (2007): "The Intersection of Age and Gender, reworking Gender Theory and Social Gerontology." In: Current Sociology 55/2, pp. 155-171.

Krotoski, A. (2004): "Chicks and Joysticks: an exploration of women and gaming." White Paper, ELSPA. Retrieved from http://cs.lamar.edu/faculty/osborne/COSC1172/elspawhitepaper3.pdf, accessed June 2015

Laurel, B. (2001): Utopian Entrepreneur, Cambridge, Massachusetts: MIT Press.

Lemon, B.W./Bengtson, V.L./Peterson, J. A. (1981): "An exploration of the activity theory of aging: Activity types and life satisfaction among in-movers to a retirement community." In C. S. Kart & B. B. Manard (Eds.), Aging in America: Readings in social gerontology (pp. 15-38). Palo Alto, CA: Mayfield. (Original work published 1972)

Lin, M.-C./Hummert, M.L./Harwood, J. (2004): "Representation of age identities in on-line discourse." In: Journal of Aging Studies 18, pp. 261-274.

Marston, H.R. (2010): "Wii Like to Play Too: Computer Gaming Habits of Older Adults." Unpublished PhD Thesis, undertaken at Teesside University, Middlesbrough, England, UK.

Marston, H.R. (2013a): "Design recommendations for digital game design within an ageing society." In: International Journal of Educational Gerontology 39/2, pp. 103-118.

Marston, H.R. (2013b): "Digital Gaming Perspectives of Older Adults: Content vs Interaction" In: Educational Gerontology 39/3, pp. 14-208.

Marston, H.R./Smith, S.T. (2012): "Interactive videogame technologies to support independence in the elderly: A narrative review." In: Games for Health Journal 1/2, pp. 139-152.

Marston, H.R. (2012): "Older adults as 21st century game designers." In: Computer Games Journal, Whitsun. Retrieved from http://tcjg.weebly.com/marston.html, accessed June 2015.

Metacritic, Video game review site, http://www.metacritic.com/games. Last accessed 2013.

Miller, K.J./Adair, B.S./Pearce, A.J./Said, C.M./Ozanne, E./Morris, M.M. (2013): "Effectiveness and feasibility of virtual reality and gaming system use at home by older adults for enabling physical activity to improve health related domains: a systematic review." In: Age and Ageing 43/2, pp. 188-195.

Morgen, L.A/ Kunkle, S. (1998). Aging: The Social Context. Sage Publications.

Neugarten B.L./Hagestad, G.O. (1976): "Age and the life course." In: Binstock, R/Shanas, E (eds.), Handbook of aging and the social sciences, New York: Van Nostrand Rein-hold, pp. 35-55.

Orlov, L.M. (2011): "Technology Survey Age 65 to 100, Extending Technology Past the Boomers." Linkage™. Retrieved from http://www.linkage-connect.com/files/1/Articles/TechnologySurveyFinalCopyFeb2012.pdf, accessed June 2015

Paoletti, I. (1998): Being an older woman: A study in the social production of identity, Mahwah, New Jersey: Lawrence Erlbaum Associates.

Pearce, C. (2008): "The truth about baby boomer gamers: a study of over-forty computer game players." In: Games and Culture 3/2, pp. 142-174.

Pratchett, R. (2005): "Gamers in the UK, Digital play, Digital Lifestyles." Commissioned by the BBC New Media & Technology: Creative Research and Development department. Retrieved from http://crystaltips.typepad.com/wonderland/files/bbc_uk_games_research_2005.pdf, accessed June 2015.

Reinharz, S. (1986): "Friends or foes: Gerontological and feminist theory." In: Women's Studies International Forum 9, pp. 503-514.

Sheffield, B. (2008): "Casual games are "A game intended for people for whom gaming is not a primary area of interest." GDC Casual Summit: Blue Fang's Meretzky Defines Casual Games. http://www.gamasutra.com/view/news/17443/GDC_Casual_Summit_Blue_Fang, accessed 05 May, 2015.

Simon, B. (2009): "Wii are out of Control: Bodies, Game Screens and the Production of Gestural Excess." Available at SSRN: http://papers.ssrn.com/sol3/papers.cfm?abstract_id=1354043, accessed June 2015.

Taylor, T.L. (2003): "Multiple Pleasures: Women and Online Gaming." In: Convergence 9/1, pp. 21-46.

Taylor, T.L. (2006): "Play between worlds: Exploring online game culture." Cambridge, Massachusetts: The MIT Press.

Turkle, S. (1984): The Second Self: Computers and the Human Spirit, Cambridge, MA, The MIT Press.

Voida, A./Carpendale, S./Greenberg, S. (2010): "The individual and the group in console gaming." In Proceedings of the ACM Conference on Computer-Supported Cooperative Work (CSCW 2010). Savannah, GA, February 6–10. New York: ACM Press, pp. 371–380.

Voida, A./Greenberg, S. (2009): "Wii all play: The console game as a computational meeting place." In Proceedings of the ACM SIGCHI Conference on Human Factors in Computing Systems (CHI 2009). Boston, Massachusetts, April 4-9. New York: ACM Press, pp. 1559-1568.

Winn, J./Heeter, C. (2009): "Gaming, gender, and time: Who makes time to play." In: Sex Roles 61/1, pp. 1-13.

Social Inclusion of Elderly People in Rural Areas by Social and Technological Mechanisms

PETER BINIOK, IRIS MENKE, STEFAN SELKE

1. INTRODUCTION

Is there a need to enhance the social inclusion of the elderly by means of technology? Is the use of contemporary information and communication technologies a way to improve the life of the elderly in rural regions? If so, how can suitable devices, platforms and software be integrated into everyday life in order to facilitate communication, contacts and exchange of information among senior citizens? These are pressing questions arising against the background of the steadily increasing senior population, its impact on society, and related disparities between urban and rural areas.

Our research, part of the joint project "SONIA – social inclusion by communication devices in urban-rural comparison" (2013-2015), addresses these questions. Furtwangen University is located in the Southwest of Germany, 1000 meters above sea-level, in a remote region of the Black Forest. Traditionally, there is an intensive interchange between the University and its region based on an ongoing culture of technical and social innovation. This lends itself to investigate the specific conditions of the rural population as a so called 'reality lab' (cf. Groß/Hoffmann-Riem/Krohn 2005; Schneidewind/ Scheck 2013) in which the dialogue between researchers and citizens is intensified systematically.[1] Methodologically, we additionally apply qualitative research methods combining semi-structured interviews with focus group discussions and document analysis, to capture both individual biographies

---

1 | By mentioning the male function function designation in this text, always both sexes are meant if not indicated differently.

(self-perception) and the participants' living conditions (outside perspective).

In our paper we focus on the issue of improving the quality of life of the elderly in rural regions. Thereby we strengthen arguments about aspects of autonomy and the enhancement of social inclusion. In this respect, we analyze how information and communication technologies may be utilized to enrich human interaction in real spaces. We thus seek to illuminate the relation between aging and technology by highlighting the valued dimensions of communication, identity and interpersonal contacts. Below we sketch out the results of our initial needs assessment and the ideas concerning our online platform concept that presents a "space of exchange" (*Raum des Austauschs*) for the elderly. This virtual space is understood as a catalyst to enable seniors and other people to get in contact and, consequently, to interact in real life places.

By the given example of our research, we finally discuss how communicative cross-linkage in virtual space revitalizes existing and establishes new personal contacts and hence fosters social inclusion. We argue that the enrichment of routinized (habitual) practices with new computational devices extends the scope of action of individuals and therefore their social inclusion. This, however, could have been expected. Unexpectedly, we found remarkable effects before the implementation of technology which we call 'paradox intervention': considering the effects of this paradox intervention, our approach questions the often stated overall increase of autonomy by assistive technologies. Taking the intervention effects into account, we introduce the concept of 'contingent autonomy growth' as a complementary perspective towards the use and effectiveness of assistive technologies.

2. Social Inclusion, the Elderly, and Rural Regions

Our study is based on the assumption that social inclusion is one dimension of 'quality of life' (cf. Nussbaum/Sen 1993; Beck et al. 1997; Nussbaum 1999). In its broad sense, social inclusion refers to societal participation of individuals achieved by interacting with other individuals. Characterized by processes of self-determination, affiliation and integration, this form of participation strongly depends on the availability of specific resources and means (cf. Wansing 2005; Lenz 2007; Schütte 2012).

Resources and Means for Inclusion

The theoretical background of our study is on the one hand informed by Bourdieus (1987) concept of types of capital (social, economic, cultural, symbolic), supplemented by the idea of corporal capital (cf. Schroeter 2005), to cover individual factors of living conditions. On the other hand, Läpples (1991) concept of societal space complements our view to make sense of specific enabling and constraining circumstances of daily life: societal practices (e.g. meeting friends), the material substrate or topography (e.g. hills, long winter), and the normative regulation system (e.g. membership in clubs and associations), as well as the system of symbols and representations (e.g. articles about the elderly in local newspapers). The various individual and structural determinants take effect as nested arrangements more or less directly on the personal, interactive, organizational and societal level (cf. Schroeter 2005).

In this way we conceptualize social inclusion as a phenomenon depending on both individual resources *and* the societal framework:[2] their combination constitutes a social environment (*Sozialraum*) people live and interact in. Social environments are understood not only as territories but as spaces of action –, so by focusing on the elderly population we assess that a "social ecology of age(ing)"[3] (Wahl 2002: 50) evolves. The more resources are available in the environment, the broader is the course of action, and the higher is the social inclusion of the inhabitants. That means that participation in everyday life requires the mobilization and transformation of appropriate means and is achieved at least by resource adjustment within the several ecologies.[4] It is important to note that social inclusion is a societal practice itself and underlies specific dynamics (cf. Wansing 2005). Social inclusion is accomplished by allocation, distribution, yet also by withdrawal of resources. Furthermore, it depends on individual biographies due to the accumu-

---

2 | A similar approach refers to lifestyle (*Lebensführung*) and conditions of life (*Lebenslage*) (cf. Backes et al. 2004; Backes/Clemens 2013; in addition see Hurrelmann/Bründel 2003) representing the two dimensions that affect social inclusion. Obviously action as well as structure (cf. Strauss 1993) has to be addressed simultaneously to modify the state of inclusion and participation.

3 | This and other quotes we translated from the original German texts.

4 | Cf. Keating (2008) for the question of the 'best fit' between elderly people and rural contexts.

lation and reinforcement of including as well as excluding factors (cf. Heusinger 2008). Following Läpple (1992), the social environment is not only something surrounding human actors – it traverses their very corporeality, their interactions, their forms of expression and ways to fulfill themselves.[5]

In the light of the demographic change (cf. Tews 1999), it is of special interest to us to observe how intensely elderly people participate in societal activities as well as to examine how participation may be increased. Therefore, the overall aim of our research is to find out the multiple ways to generate and/or re-activate possibilities to increase the number of actions and interactions of the elderly. What information is needed to stimulate interactions? How can opportunities for leisure activities be distributed? What is of special interest within a social environment?

Aging and the Rural

There is no single category of "age" as age-dependent classification, as the group of people older than 60 years is very heterogeneous.[6] This heterogeneity results from different lifestyles and circumstances as well as from a wide range of social inequality patterns, correlating with previous educational and employment opportunities (cf. VDE 2008). On the one hand, age and aging may be associated with positive attributes like gaining serenity and foresight, and a rise in the quality of life (cf. Jakobs et al. 2008). On the other hand, age encompasses rather negative aspects, such as physical degeneration and social exclusion (ibid.). Studies on aging mostly target urban regions and have a limited focus on core themes, as for instance on living conditions concentrating on habitation and quarters (cf. Oswald 2002; Büscher et al. 2009; Kaiser 2012), or on issues of care and health protection (cf. Heusinger 2008; Walter/Schneider 2008). More often than not, "aging in rural regions is frequently idyllic transfigured or negatively exaggerated" (Walter/Altgeld

---

5 | Cf. Gieryn 2000 for a place-sensitive sociology and Löw 2001 for a concept of space that integrates action and structure.

6 | Staudinger and Schindler (2002) for example suggest distinguishing at least three phases of age and aging: young elderly in their sixties, elderly between 70 and 85 years old, and old and very old people over 85.

2002: 318) and studies that explicate on aging in rural regions more appropriately remain underrepresented.[7]

There are various differences between urban and rural regions (cf. Walter/Altgeld 2002; Wahl/Schilling/Oswald 2002; Scherger/Brauer/Künemund 2004; Bohl 2005). Some central characteristics of the rural are: low density of population (with about 100 inhabitants per km^2), decrease of health care in line with population and centrality, infrastructure as crucial problem, and a low range of leisure activities and other opportunity structures.[8] In general, families have a higher number of children and large households with commonly property ownership and long periods of residence. Distances between generations of one family are low. Additionally, one finds a strong integration in local structures and support systems as well as normative values and high social control hampering the acceptance of professional care. These aspects of the social environment have been commonly expressed by our interviewees when asked about their needs and wishes (see also section 4).

To recapitulate, our research follows the question of how social inclusion of the elderly population may be fostered with regard to the modification of the individual resources (types of capital) *and* the transformation of societal framework conditions. While the elderly are able to configure their activities relative to their life styles (cf. Kolland/Rosenmayr 2007), informal ties and contacts especially have the most positive effects on their life satisfaction. Yet, the elderly need help and assistance on different levels and with regard to their social, material and technological environment, e.g. administer medicine, housekeeping or companionship (cf. Isfort 2013). In this respect, we also examine how information and communication technologies may be utilized to enrich human interaction and communication. The question of how the quality of life of elderly persons can be enhanced by using computers, tablets or smart phones is investigated. In this, we seek to contribute to the research field of assistive technologies, which increasingly receives public funds under the 'umbrella term' (Rip/Voß 2013) "Ambient Assisted Living".

---

7 | There are several exceptions that steadily fill this gap, e.g. Schilling/Wahl 2002, Mollenkopf/Kaspar 2005, Keating 2008, Hofreuter-Gätgens et al. 2011, Hennessy/Means/Burholt 2014.

8 | For further discussions on formal indicators of rural regions see Bohl 2005, Neu 2010 or Penke 2012.

3. Ambient Assisted Living: Funding, Baselines and Expectations

Ambient Assisted Living (AAL) refers to "living in an environment supported by 'intelligent' technologies, which react sensitive and adaptive on the presence of humans and objects and fulfill various services" (Becks et al. 2007: 3). The enfolding field AAL "includes concepts, products and services, which increase and secure live quality by use of information and communication technologies" (Georgieff 2008: 23). Ambient and assistive technologies are in this perspective the 'good folk' (cf. Rammert 1998) that hide and come up if needed (assistive), and otherwise monitor themselves and human actions and interactions in the background (ambient).

Thinking about Aging

Scientific research on AAL is nowadays intensively funded by the German government and thus framed by political-normative assumptions. The discourse of demographic change is the starting point for different initiatives on the national level, as for example the research agenda "age has future" (*Das Alter hat Zukunft*) or the main funding area "man machine interaction within demographic change" (*Mensch-Technik-Interaktion im demografischen Wandel*)[9] (cf. BMBF/VDE 2011). These efforts go hand in hand with an increasing public interest in aging and with shifts in our understanding of what age means (cf. Denninger et al. 2014 for the detailed description of this change; cf. Pauly 1995 for the general approach of 'shifting baselines').[10] This development leads Reindl (2009: 168) to state: "The prevalent image of age and aging predetermines stereotypes that differ heavily from traditional interpretations of age and aging."

---

9 | Cf. http://www.das-alter-hat-zukunft.de and http://www.mtidw.de.

10 | These efforts grow out from the changing role of the nation state. Following a critical perspective (cf. Denninger et al. 2014) one may scrutinize, whether or not the state slowly abdicates liability towards its citizens and transfers responsibility on individuals, technologies ('adiaphorization' as stated in Bauman/Lyon 2013) and local solidarity networks and liability communities (*Verantwortungsgemeinschaften*). Simultaneously in all areas of society the responsible self is proclaimed, be it as 'entreployee' (*Arbeitskraftunternehmer*), in health protection or pension planning.

As opposed to the 1980s notion of earned retirement and withdrawal from active live (known as 'disengagement'), viewpoints of productive and active aging predominate today (cf. in general WHO 2002; taken up for instance in Walker 2002). Politics specially promote volunteer work and civic engagement to use and increase the potential of age, which is in line with the postulation to increase the social participation of the elderly population. This can be found, for instance, in the aging report from the German Government: "Against the background of demographic change, the [German] Government pursues the goal to *foster the potential of engagement of the elderly*. In our society of long life it is important to stay active as long as possible, and to participate, to commit oneself, be it in work life, in civic engagement, or in family" (Deutscher Bundestag 2010: VII, emphasis added). The aging population is seen as a fit and adventurous population, consisting of 'silver surfers' and 'best agers'. Through assistive technologies, the 'activation society' (*Aktivierungsgesellschaft*) offers the necessary means to achieve more autonomy and self-determination and the underlying paradigm of the active elderly manifests itself in the technology-prospective framework of AAL.

Technology as a Resource

With regard to the afore-mentioned interdependence of resources and social inclusion, we identify a technology-deterministic argument for the use of technologies within the field of AAL (Georgieff 2009: 26, emphasis added; see also Lancioni/Singh 2014): "Technologies have to be seen as a *crucial resource of the environment of the elderly*. On the one hand they help to compensate capacity losses and disabilities and on the other hand technologies optimize life quality and enrich day to day aging." Consequently, technologies as a resource will be of increasing importance in the future. Technologies are seen as suitable devices to support aging and caring, while simultaneously implying market potential and the saving of costs, e.g. in the health care system (cf. Georgieff 2008; VDE 2008; Wichert/Norgall 2009). The resource technology needs to be used to relieve, for instance, caring relatives or (mobile) nursing services. Beyond that, information and communication technologies are probable means to increase autonomy and security of the elderly themselves. Smart homes and smart living represent the ideal of staying home for lifetime – whether healthy, ill or otherwise care-dependent.

Such a technology-driven (if not deterministic) perspective mostly neglects the human factor, e.g. the wish to disengage, a lack of societal structures to use technologies meaningfully, or missing competences in the usage of information and communication technologies (cf. Pelizäus-Hoffmeister 2013). Moreover, concepts of AAL often oversee that the elderly are not a homogeneous cohort or group and a variety of engineered lifestyles exist (cf. Mollenkopf 2006). Accordingly, it will be difficult to develop age-appropriate technologies (cf. Hogreve et al. 2011) or 'personalized technologies' (as, for instance, within the upcoming approach of 'personalized medicine'). Lastly, studies on effects and utilization of existing assistive technologies hardly exist (cf. Brandt et al. 2011; Anttila et al. 2012).

Regional (Funding) Contexts

Apart from funding and research agendas at the German national level, federal states like Baden-Wuerttemberg also position themselves as proponents of research on aging and care, while strongly supporting programs and projects on the regional level. One of these efforts following the current image of age and aging is the impulse program "Medicine and Care" from the Ministry of Labor and Social Affairs, Families, Women, and Senior Citizens.[11] Starting from the paradigm "outpatient instead of inpatient", concepts of technology-based maintenance are being developed to enable diseased and old people to stay at home and to support caring services.

Within this program, the joint project "SONIA – social inclusion by communication devices in urban-rural comparison" is funded. One of the aims in this project is to investigate the social inclusion of people older than 60 years in rural and urban regions. Due to the potential need to increase social inclusion, we further determine the adaption of existing information and communication technologies to serve these needs. SONIA combines multiple disciplinary and scientific perspectives and approaches on the development of assistive technologies. In the following sections, we present the *social-scientific point of view* and our preliminary findings from the rural case study. The rural region in our project is exemplified by the village Furtwangen and its adjoining municipalities, where Furtwangen University is located and where we conduct our research. This focus is not (only) due to practical reasons: rather, the major feedback on our efforts to encourage people to participate in the project came

---

11 | The program is carried out in cooperation with the Ministry of Finances and Economy as well as the Ministry for Science and Art.

from that region. Besides that, during our research we realized that our concept of social inclusion has to be connected to local structures and actions.

4. Needs, "Space of Exchange" and the 'Paradox of Intervention'

With our user-centered approach we met the challenge to put human beings in the first place instead of technology: we started our research with a needs assessment to learn about the wishes and demands of the elderly.[12] Our needs assessment combines qualitative methods, such as personal narrative interviews to learn about the biography and living environment, focus group discussions, and document analysis (cf. Schütze 1983; Hopf 2005). The results of the needs assessment were used to develop and implement a communication concept called "space of exchange". This online platform is tested by the elderly, which in turn allows us to analyze its impact on social inclusion.

Getting into Conversation

The target group for the interviews has been chosen following specified criteria to ensure a broad variety in terms of personal networks, monetary resources and mobility (meaning both physical fitness and cognitive abilities). These central dimensions of comparison are regarded as indicators of risk and protection respectively and are frequently used in similar studies (cf. Mollenkopf et al. 2001; Oswald 2003; Mollenkopf/Kaspar 2005). To recruit the elderly, we organized workshops, launched ads in local newspapers and printed flyers. In addition, we made use of several intermediators to get in contact with our target group, e.g. the local domestic nursing services. Besides this, over the course of the interviews we made use of snowball sampling. Altogether, we conducted 26 interviews with 16 men and ten women,

---

12 | Similar approaches as for instance 'scenario-based design' one can find in Compagna et al. (2011) and Cieslik et al. (2012). For the general idea of participatory technology development see Giesecke (2003) or Bieber/Schwarz (2011). However, technology-based approaches differ in their methodological orientation and sometimes disregard social-scientific concepts and methods (cf. Pelizäus-Hoffmeister 2013).

mostly between 70 and 80 years old and respective to available resources on average in good constitution.[13]

As for the data collection, apart from the personal interviews with the primary users (the elderly), several workshops and focus group discussions with secondary users (representatives of local nursing services, clubs and societies) were conducted, as they play a central role concerning the implementation of technological devices to enhance the social inclusion of the elderly. After we presented them the results from the interviews, they were asked as experts to discuss and conceptualize the virtual "space of exchange" which in the end is supposed to lead to social inclusion in real life.

Necessity Matrix and Central Challenges

Our notion of aging in rural regions is formulated under usage of a *necessity matrix* that will only be sketched at this point. Three core dimensions generate the matrix: "interactive self-organization" as the wish for autonomy and the independent arrangement of day structure and leisure activities, "intergenerational relationships" and "interactivity radius" (see Table 1).

Interactive self-organization includes possibilities to (inter)act independently and shows four characteristics: obligations are not negotiable (as care roles), periodicals are important but not necessary (shopping, group of regulars), options are at choice (a cup of coffee in the neighborhood, go for a stroll, go to the movies), and coincidence means single or irregular incidents (traveling acquaintance). Relevant groups of agents were pointed at in the interviews and actions of divergent autonomy correspond to these agents. As a result, a network of relationships linked to specific activities reveals social, cultural and cognitive distances; this is represented in the dimension *intergenerational relationships*. The constellation of agents concerning activities shows six different manifestations: alone, with partner, with family (without partner), with "the young" (field-specific term for the younger generation), with the same generation, and with strangers. The territorial space, in which

---

13 | The interviews were transcribed and analyzed with the aid of the qualitative data analysis software MAXQDA, following a category system developed on the basis of the core statements. Using 'thematic coding', enhanced by the 'open coding' approach (cf. Strauss/Corbin 1990; Schmidt 1997, 2005; Kuckarzt 2010), we found correlations that lead us to the development of a consistent notion of aging in rural regions.

activities are performed, comprises the more or less well known surroundings of the interviewee. This space to which the range of activities relates is reflected in *interactivity radius*. Concerning the range of activities, five manifestations can be determined: in the house or flat, immediate neighborhood, in the neighborhood, in the village, outside the village.

Crossing these three dimensions generates a variety of arrays. The arrays represent typical everyday situations which are, regarding social inclusion, satisfied, showing up conflicts or have not yet occurred. In this way the matrix serves as a useful heuristic tool to identify daily challenges and characteristics of aging in rural regions. A description of specific situations to be improved or even realized derived from the empirical data. They have been determined and lead to the identification of the needs of the elderly. Moreover, the determination of needs is interconnected to individual and contextual factors. On an abstract level, we identified three significant wishes: the wish for safety by horizons of possibilities, the wish for everyday life without obligations, and the wish for socializing despite disengagement.

In more detail, central needs and wishes of the senior citizens in rural regions correspond to the typical features of the Black Forest's landscape, its long winter and agriculture (see section 2). This influences, first of all, *mobility*: being mobile (both at home and outside), traveling around, and driving a car (cf. Mollenkopf/Kaspar 2005; Jakobs et al. 2008). In rural regions it is unavoidable to drive a car to get to doctors, to theatres or to shopping, because the local transport services run very rarely. If elderly persons are unable to drive, they become dependent upon their families or neighbors or service providers. The same is true in winters with deep snow and iced-up hillsides. Therefore, and secondly, *supportive services* (e.g. delivery services, medical support) are of importance. Concerning this matter, the impact of family structures has become evident in our study (cf. Schilling/Wahl 2002; Jakobs et al. 2008) together with the relevance of a vivid neighborhood as local resource (cf. Rädler/Schubert 2012). This is, however, of indeterminate duration: the same-aged neighborhood will especially become problematic as resource in the future. Thirdly, the need for the *work of associations and leisure activities* (as hiking and sociability) is stressed (cf. Baumgartner et al. 2013). Here it has to be remarked that in many towns and villages associations are faced with membership decline and the problem to find new chairmen.

Table 1: Necessity Matrix

[2] Intergenerational relationships	[1] Interactive self-organization				[3] Interactivity radius
	obligations	periodicals	options	coincidence	H – house/flat I – immediate neighborhood N – neighborhood V – village O – outside village
alone					H I N V O
with partner					H I N V O
with family					see above
with "the youth"					
with the same generation					
with strangers					

Conceptualization: From Needs to Apps

From this collection of wishes and needs, we firstly extracted those relatively general issues that could be addressed by strengthening communication (such as the motivation to go out for a walk or support networks in the village). Further, we selected needs that could be satisfied with information and communication technologies: establishing new personal ties, information on (medical) services, support associations, diverting social contacts, and others. After identifying these wishes and needs, a concept for technical support was developed and subsequently put into practice. Our aim was to meet the needs in accordance and compliance with already existing regional social and organizational structures (such as community college, local foundations, clubs and associations). For that reason, we focused our research activities on the existing structures in the region Furtwangen (Black Forest).

The core of our conceptual framework is a "space of exchange" that is carried out as an online platform with several applications grouped under thematic topics with indigenous terminologies (see Figure 1). The elderly access the "space of exchange" with tablet computers. There we provide, for example, a calendar to broadcast association activities ("Wa isch los im Schtädtli?"), information about the local civic-mobile or local news ("Wunderfitz hät d Nase gschpitzt"), a black board to exchange assistance ("Am Schwarze Brätt") and a chat function to make appointments or just to jabber ("Schwätze"). This virtual space exceeds provision and exchange of information. It also facilitates entertainment, the launching of discussions as well as mutual supply of and demands for assistance. The overall objective is to stimulate and foster interactions in *real life*.

We consider our platform to be distinctive from other forms of assistive technologies for three reasons: (1) social inclusion operates as self-enabling social practices of participation instead of externally governed integration mechanisms, (2) our platform is assistive in terms of an interactive communication portal – not a technology in terms of a reactive monitoring and information system, and (3) technology is a means or a medium that does not substitute social contacts and face-to-face interactions. To have our platform tested by the elderly and to investigate its effects on their life and on social inclusion is the current step of our project.

Figure 1: "Space of Exchange" Start Screen[14]

'Paradox of Intervention' – Re-structuring the Social Landscape by Research

To get in contact with the rural population has been our initial activity. We interviewed them at home, we talked to them on the telephone, and we met them at workshops and information events. Interestingly, we observed a lot of engagement among the elderly even before we deployed the technology. Since our project seems to stimulate effects based on social and not technological mechanisms, we describe these foregoing activities from the project's perspective as a 'paradox of intervention'. In general, it is obvious that attention on the part of (social) sciences is a positive irritation – a welcome change in day to day routine, especially for the elderly. The engagement of our participants can be described as follows.

With regard to our interviews, we *first* notice only a slight reflection on aging. Our research now opens up the opportunity for seniors to deal with the issue "good life in rural regions". The questions we ask, – such as provision for old age, neighborhood, or sacrifice and loss – stimulate our dialogue partners to think more intensely about their contentment and position in society. Moreover, our workshops are the locus for the re-generation of ideas and

---

[14] | The underlying technology ("CareBW") has been developed by the company nubedian (Karlsruhe, Germany).

the recall of past hobbies or leisure activities. One woman comments on the black board of our platform and has the consideration to invite people on the occasion of her baking a cake – something she loves to do but has not done for years. Others for example think about sharing a car to go to the doctor.

Secondly, the participation in our study is understood by the elderly as self-commitment. Apart from their continuous interest and curiosity, the participants stake out claims. After one of the workshops, an angered woman inquires the following morning why she had not been invited while her friend has been: she stresses to have been among the first persons to have contributed to our project as interviewee. This is an example of how our research influences expectations and behavior of our participants. Their self-commitment goes hand in hand with the assumption that we as researchers commit ourselves to the task to enhance the life quality of the elderly population – so some elderly voiced their displeasure if we were not in contact with them for weeks. Of course there are more motives to join our project than that, as for instance keeping up with new technologies or an interest in scientific research per se, but to be taken seriously as citizens seems to be a crucial factor to participate.

Thirdly, we notice a steady creation and re-activation of personal contacts and strengthening of networks. That is especially fostered by our meetings. As one participant stated, "We never met in this constellation – this is fantastic". To run our platform the characteristics of the locality Furtwangen require the establishment of different teams for administration, training courses and support.[15] Here we can access volunteers who support our research. Therefore, with our project, we structure the social landscape as net-workers or net-builders more or less intentionally. Such meetings and informal conversations between researchers and population generate communication on a high qualitative level with regard to the solution of societal problems and is not mere an increase in number.

This 'paradox intervention' is a result of our research, but was not intended. Following our research plan, the intervention begins with the test of our "space of exchange", but now the question arises: What does social inclusion (yet) depend on?

---

15 | This is different from the parallel sub-project concerned with a quarter in an urban region where a central management and community center exist.

5. Conclusions: Social Inclusion by 'Man' or 'Machine'?

Is there a need to use information and communication technologies to enhance the social inclusion of the elderly in rural areas? This summarizing question can be answered as follows. Information and communication technologies do not need to be applied, but can be applied and work great as mediators. Starting from the assumption that the elderly are restricted in their activities and gain access only to a little course of actions and small 'opportunity structures' (Merton 1996), ambient assisted technologies are intended to increase such possibilities and thereby increase autonomy too. In table 2 this process can be thought of as a move from box (1) to box (3).

As we have shown, on this *ostentatious* level a wide range of normative mechanisms take effect on scientific and technology development (see section 3). The term ostentatious refers to the discursive level of scientific practice. In our view the postulate of autonomy is demonstrative and purposefully put on display. A lot of projects are funded to investigate the darker sides of aging and how the elderly population can be supported and assisted. We argue that the promises associated with technology are often overstated or stated too universally, and that the enhancement of autonomy and self-determination is limited or do not necessarily take place, as clarified by box (4) in table 2. Inside the field of AAL, autonomy and safety are usually central ideas that are much claimed and demanded but less challenged (cf. Selke/Biniok 2015). There are barely experiences and differentiations regarding the effects and benefits of assistive technologies.[16]

We implemented our assistive technology and observed effects on the day to day life of the elderly on a *performative* level of scientific practice. In difference to asserted (ostentatious) effects, initial changes unfold with regard to action and interaction. In table 2, this is expressed by a move from box (1) to box (2). There seem to be social mechanisms and rules that enhance social en-

---

16 | Correspondingly, we experience a skeptical attitude towards our research. Criticism is expressed especially with regard to missing structures (organizer, moderator, and 'troubleshooter') on behalf of the public and the economic interests related with our project and with the implementation of an assistive technology. Without the necessary social structures, our project is considered as an unsuccessful endeavor, but this prejudice can be read as a criticism of AAL and its projects in general.

gagement and inclusion, so we have the paradoxical situation of an intervention before the intervention originally intended by the project. This paradox is not new: two lines of reasoning beyond placebo or bias can be suggested.

Table 2: Social Inclusion and Intervention

	Many opportunities and high autonomy	Few opportunities and low autonomy
Social mechanisms	(2) Increase by research and/or "by chance"	(1) Demand for investigation and intervention
Technological mechanisms	(3) Increase by AAL	(4) No or limited effect by AAL

From a socio-psychological point of view, our activities with their orientation towards the social environment follow an approach (cf. Albrecht 2008) known as 'social environment pastoral' (*Sozialraumpastoral* or *sozialraumorientierte Pastoral*). The idea behind this approach is to strengthen local or regional communities by paying attention to the needs and wishes of the citizens, using available resources and connecting various actors. That means the altruistic engagement of third parties – e.g. churches – can help people to feel better and/or become more active: "Care is perceived as relief and enrichment in many cases only by its presence." (Wingenfield/Schaeffer 2001: 144) Even though pastoral acting results from religious objectives, these motivations fade from the spotlight in daily practice. In a similar way, as we show interest in the life of the elderly in rural areas, and take them seriously, we evoke re-actions by perceived support.[17]

In a more scientific and methodological perspective, we bring about a 'Hawthorne effect' (Roethlisberger/Dickson 1939) on the one hand and some kind of 'serendipity of science' (Merton/Barber 2006) on the other. The former effect refers to the fact that a mere participation in an experiment or

---

17 | Social work (*soziale Arbeit*) follows similar objectives and orients itself towards social environments (cf. Debiel et al. 2012).

a study affects the behavior of the participants.[18] Again, it is the degree of attention that leads to actions and interactions, which in turn give rise to new personal networks, new alliances and modifications of living conditions. This observation of unanticipated results during our research is what Merton and Barber call serendipity. Although the structure of our study presets it differently, we found by chance social factors influencing social inclusion. The paradox intervention before intervention is now the basis of our future research and serves as a model to deepen our understanding of social inclusion influenced by social *and* technological determinants.

That points to the potential of assistive technologies. In our view autonomy (or self-determination and independence) grows out especially from available courses of action within the living environment: multiplication of opportunities of action generates autonomy. These opportunities are usually limited, only limited expendable, and emerge only under specific conditions and with regard to specific social groups and timescales. For that reason, we suggest speaking of 'contingent autonomy growth' instead of the normative demand for autonomy. At present we feel that empowerment in its ideological manifestation is misleading. We argue that the delegation of responsibilities to individual citizens has to be linked with catalyst mechanisms, whether they are social or technical. Personal responsibility is as crucial as the transformation of societal contexts.

To spur somebody into action does not reach far enough. Instead, as our project does, an opening of possibilities for interaction has to occur. Technology, if then, has to be fit in as artifact on the structural level and as something 'handiness' (Heidegger 1927) on the practice level. *In sum*, autonomy is a fluid mosaic consisting of existing and missing opportunities in life, which obtains its dynamics from changing circumstances of life and modified ways of living. Technologies only should be inserted and applied if there is more knowledge of dependencies, specific needs and wishes, and living environments.

---

18 | Already Thomas and Thomas (1928) point out: "If men define situations as real they are real in their consequences."

Acknowledgements

The project is funded by the Ministry of Labor and Social Affairs, Families, Women, and Senior Citizens Baden-Wuerttemberg, Germany. We thank our project partners at Furtwangen University, the Fraunhofer Institute for Industrial Engineering (IAO) and the Fraunhofer Institute for Systems and Innovation Research (ISI), at the University Hospital Tübingen as well as at the Development Centre "Good Aging" Ltd. and the foundation Paul Wilhelm von Keppler for inspiring discussions. Furthermore, we are indebted to all the people who were willing to tell us about their lives and work and who continue to support our research at Furtwangen University. Lastly we are grateful to Ines Hülsmann assisting us during the translation of our manuscript.

References

Albrecht, P.-G. (2008): Professionalisierung durch Milieuaktivierung und Sozialraumorientierung? Caritas-Sozialarbeit in der Entwicklung, Wiesbaden: VS.

Anttila, H./Samuelsson, K./Salminen, A.-L./Brandt, Å. (2012): "Quality of evidence of assistive technology interventions for people with disability: An overview of systematic reviews." In: Technology and Disability 24, pp. 9-48.

Backes, G./Clemens, W. (2013): Lebensphase Alter. Eine Einführung in die sozialwissenschaftliche Alternsforschung, Weinheim: Juventa.

Backes, G./Clemens, W./Künemund, H. (eds.) (2004): Lebensformen und Lebensführung im Alter, Wiesbaden: VS.

Bauman, Z./Lyon, D. (2013): Daten, Drohnen, Disziplin. Ein Gespräch über flüchtige Überwachung, Berlin: Suhrkamp.

Baumgartner, K./Kolland, F./Wanka, A. (2013): Altern im ländlichen Raum. Entwicklungsmöglichkeiten und Teilhabepotentiale. Stuttgart: Kohlhammer.

Beck, W./van der Maesen, L.J.G./Walker, A. (eds.) (1997): The Social Quality of Europe, The Hague, London, Boston: Kluwer Law International.

Becks, T./Dehm, J./Eberhardt, B. (2007): Ambient Assisted Living. Neue „intelligente" Assistenzsysteme für Prävention, Homecare und Pflege, Frankfurt am Main: DGBMT im VDE.

Bieber, D./Schwarz, K. (eds.) (2011): Mit AAL-Dienstleistungen altern. Nutzerbedarfsanalysen im Kontext des Ambient Assisted Living, Saarbrükken: iso.

BMBF/VDE Innovationspartnerschaft AAL (eds.) (2011): Ambient Assisted Living (AAL) – Komponenten, Projekte, Services. Eine Bestandsaufnahme, Berlin: VDE-Verlag.

Bohl, K.F. (2005): "Sozialstruktur." In: Beetz, S./Brauer, K./Neu, C. (eds.), Handwörterbuch zur ländlichen Gesellschaft in Deutschland, Wiesbaden: VS, pp. 225-233.

Bourdieu, P. (1987): Sozialer Sinn. Kritik der theoretischen Vernunft, Frankfurt am Main: Suhrkamp.

Brandt, Å./Samuelsson, K./Töytäri, O./Salminen, A-L. (2011): "Activity and participation, quality of life and user satisfaction outcomes of environmental control systems and smart home technology: a systematic review." In: Disability & Rehabilitation: Assistive Technology 6/3, pp. 189-206.

Büscher, A./Emmert, S./Hurrelmann, K. (2009): "Die Wohnvorstellungen von Menschen verschiedener Altersgruppen" In: Veröffentlichungsreihe des Instituts für Pflegewissenschaft an der Universität Bielefeld (IPW), Bielefeld: Institut für Pflegewissenschaft.

Cieslik, S./Klein, P./Compagna, D./Shire, K.A. (2012): "Das Szenariobasierte Design als Instrument für eine partizipative Technikentwicklung im Pflegedienstleistungssektor." In: Shire, K.A./Leimeister, J.M. (eds.), Technologiegestützte Dienstleistungsinnovation in der Gesundheitswirtschaft, Wiesbaden: Gabler, pp. 85-110.

Compagna, D./Derpmann, S./Helbig, T./Shire, K.A. (2011): "Partizipationsbereitschaft und -ermöglichung einer besonderen Nutzergruppe. Funktional-Partizipative Technikentwicklung im Pflegesektor" In: Bieber, D./Schwarz, K. (eds.), Mit AAL-Dienstleistungen altern. Nutzerbedarfsanalysen im Kontext des Ambient Assisted Living, Saarbrücken: iso, pp. 161-176.

Debiel, S./Engel, A./Hermann-Stietz, I./Litges, G./Penke, S./Wagner, L. (eds.) (2012): Soziale Arbeit in ländlichen Räumen, Wiesbaden: VS.

Denninger, T./van Dyk, S./Lessenich, S./Richter, A. (2013): Leben im Ruhestand. Zur Neuverhandlung des Alters in der Aktivgesellschaft, Bielefeld: transcript.

Deutscher Bundestag (ed.) (2010): Sechster Bericht zur Lage der älteren Generation in der Bundesrepublik Deutschland. Unterrichtung durch die Bundesregierung, Berlin: Deutscher Bundestag.

Georgieff, P. (2008): Ambient Assisted Living. Marktpotenziale IT-unterstützter Pflege für ein selbstbestimmtes Altern, Stuttgart: MfG Stiftung.

Georgieff, P. (2009): Aktives Alter(n) und Technik. Nutzung der Informations- und Kommunikationstechnik (IKT) zur Erhaltung und Betreuung der Gesundheit älterer Menschen zu Hause, Karlsruhe: Fraunhofer-Institut für System- und Innovationsforschung.

Gieryn, T.F. (2000): "A Space for Place in Sociology." In: Annual Review Sociology 26, pp. 463-496.

Giesecke, S. (ed.) (2003): Technikakzeptanz durch Nutzerintegration? Beiträge zur Innovations- und Technikanalyse, Teltow: VDI-Technologiezentrum.

Groß, M./Hoffmann-Riem, H./Krohn, W. (2005): Realexperimente. Ökologische Gestaltungsprozesse in der Wissensgesellschaft, Bielefeld: transcript.

Heidegger, M. (2006[1927]): Sein und Zeit, Tübingen: Max Niemeyer.

Hennessy, C.H./Means, R./Burholt, V. (eds.) (2014): Countryside Connections. Older People, community and place in rural Britain, London: Policy Press.

Heusinger, J. (2008): "Der Zusammenhang von Milieuzugehörigkeit, Selbstbestimmungschancen und Pflegeorganisation in häuslichen Pflegearrangements älterer Menschen." In: Bauer, U./Büscher, A. (eds.), Soziale Pflegeungleichheit, Wiesbaden: VS, pp. 301-314.

Hofreuter-Gätgens, K./Mnich, E./Thomas, D./Salomon, T./von dem Knesebeck, O. (2011): "Gesundheitsförderung für ältere Menschen in einer ländlichen Region." In: Bundesgesundheitsblatt 54, pp. 933-941.

Hogreve, J./Bilstein, N./Langnickel, D. (2011): "Alter schützt vor Technik nicht? Zur Akzeptanz technologischer Dienstleistungsinnovationen von Senioren." In: Bieber, D./Schwarz, K. (eds.), Mit AAL-Dienstleistungen altern. Nutzerbedarfsanalysen im Kontext des Ambient Assisted Living, Saarbrücken: iso, pp. 32-50.

Hopf, C. (2005): "Qualitative Interviews – ein Überblick." In: Flick, U./von Kardorff, E./Steinke, I. (eds.): Qualitative Forschung. Ein Handbuch, Reinbek: Rowohlt, pp. 349-360.

Hurrelmann, K./Bründel, H. (2003): Einführung in die Kindheitsforschung, Weinheim, Basel, Berlin: Beltz.

Isfort, M. (2013): "Anpassung des Pflegesektors zur Versorgung älterer Menschen." In: Aus Politik und Zeitgeschichte 63/4-5, pp. 29-35.

Jakobs, E.-M./Lehnen, K./Ziefle, M. (2008): Alter und Technik. Studie zu Technikkonzepten, Techniknutzung und Technikbewertung älterer Menschen, Aachen: Apprimus.

Kaiser, G. (2012): Vom Pflegeheim zur Hausgemeinschaft, Köln: KDA.

Keating, N.C. (ed.) (2008): Rural ageing. A good place to grow old?, London: Policy Press.

Kolland, F./Rosenmayr, L. (2007): "Altern und zielorientiertes Handeln. Zur Erweiterung der Aktivitätstheorie." In: Wahl, H.-W./Mollenkopf, H. (eds.), Alternsforschung am Beginn des 21. Jahrhunderts, Berlin: Akademie, pp. 203-221.

Kuckarzt, U. (2010): Einführung in die computergestützte Analyse qualitativer Daten, Berlin: Springer.

Lancioni, G.E./Singh, N.N. (eds.) (2014): Assistive Technologies for People with Diverse Abilities, New York: Springer.

Läpple, D. (1991): "Essay über den Raum: für ein gesellschaftswissenschaftliches Raumkonzept." In: Häussermann, H./Ipsen, D./Krämer-Badoni, T. (eds.), Stadt und Raum, Pfaffenweiler: Centaurus, pp. 157-207.

Lenz, I. (2007): Inklusionen und Exklusionen in der Globalisierung der Arbeit, Berlin: Springer.

Löw, M. (2001): Raumsoziologie, Frankfurt am Main: Suhrkamp.

Merton, R.K. (1996): On Social Structure and Science, Chicago, London: University of Chicago Press.

Merton, R.K./Barber, E. (2006). The Travels and Adventures of Serendipity: A Study in Sociological Semantics and the Sociology of Science, Princeton: University Press.

Mollenkopf, H. (2006): "Techniknutzung als Lebensstil?." In: Kimpeler, Simone und Elisabeth Baier (eds.), IT-basierte Produkte und Dienste für ältere Menschen – Nutzeranforderungen und Techniktrends, Stuttgart: Fraunhofer IRB, pp. 65-78.

Mollenkopf, H./Kaspar, R. (2005): "Ageing in rural areas of East and West Germany: increasing similarities and remaining differences." In: European Journal Aging 2, pp. 120-130.

Mollenkopf, H./Oswald, F./Schilling, O./Wahl, H.-W. (2001): "Aspekte der außerhäuslichen Mobilität älterer Menschen in der Stadt und auf dem Land: objektive Bedingungen und subjektive Bewertung." In: Sozialer Fortschritt 50, pp. 214-220.

Neu, C. (2010): "Land- und Agrarsoziologie." In: Kneer, G./Schroer, M. (eds.), Handbuch spezielle Soziologien, Wiesbaden: VS, pp. 243-261.

Nussbaum, M.C. (1999). Gerechtigkeit oder Das gute Leben, Frankfurt am Main: Suhrkamp.

Nussbaum, M.C./Sen, Amartya (1993): The Quality of Life, Oxford: Clarendon Press.

Oswald, F. (2002): "Wohnbedingungen und Wohnbedürfnisse im Alter." In: Schlag, B./Megel, K. (eds.), Mobilität und gesellschaftliche Partizipation im Alter, Stuttgart: Kohlhammer, pp. 97-115.

Oswald, F./Wahl, H.-W./Mollenkopf, H./Schilling, O. (2003): "Housing and life satisfaction of older adults in two rural regions in Germany." In: Research on Aging 25, pp. 122-143.

Pauly, D. (1995): "Anecdotes and the shifting baseline syndrome of fisheries." In: Trends in Ecology and Evolution, 10/10, p. 430.

Pelizäus-Hoffmeister, H. (2013): Zur Bedeutung von Technik im Alltag Älterer: Theorie und Empirie aus soziologischer Perspektive, Wiesbaden: Springer.

Penke, S. (2012): "Ländliche Räume und Strukturen – mehr als eine ‚Restkategorie' mit Defiziten." In: Debiel, S./Engel, A./Hermann-Stietz, I./Litges, G./Penke, S./Wagner, L. (eds.), Soziale Arbeit in ländlichen Räumen, Wiesbaden: VS, pp. 17-27.

Rädler, M./Schubert, H. (2012): Innovation im ländlichen Raum am Beispiel des Allgemeinen Sozialen Dienstes (ASD). In: Debiel, S./Engel, A./Hermann-Stietz, I./Litges, G./Penke, S./Wagner, L. (eds.), Soziale Arbeit in ländlichen Räumen, Wiesbaden: VS, pp. 109-119.

Rammert, W. (1998): "Giddens und die Gesellschaft der Heinzelmännchen." In: Malsch, T. (ed.), Sozionik, Berlin: Edition Sigma, pp. 91-128.

Reindl, J. (2009): "Abschaffung des Alters." In: Leviathan 37, pp. 160-172.

Rip, A./Voß, J.-P. (2013): "Umbrella Terms as Mediators in the Governance of emerging Science and Technology." In: Science, Technology & Innovation Studies 9/2, pp. 39-59.

Roethlisberger, F.J./Dickson, W.J. (1939): Management and the worker: an account of a research program conducted by the Western electric company, Hawthorne works, Chicago, Cambridge, Massachusetts: Harvard University Press.

Scherger, S./Brauer, K./Künemund, H. (2004): "Partizipation und Engagement älterer Menschen – Elemente der Lebensführung im Stadt-Land-Vergleich." In Backes, G./Clemens, W./Künemund, H. (eds.), Lebensformen und Lebensführung im Alter, Wiesbaden: VS, pp. 173-192.

Schilling, O./Wahl, H.-W. (2002): "Familiäre Netzwerke und Lebenszufriedenheit alter Menschen in ländlichen und urbanen Regionen." In: Kölner Zeitschrift für Soziologie und Sozialpsychologie 54/2, pp. 304-317.

Schmidt, C. (1997): "«Am Material»: Auswertungstechniken für Leitfadeninterviews." In: Barbara Friebertshäuser/Prengel, Annedore (eds.), Handbuch Qualitative Forschungsmethoden in der Erziehungswissenschaft, Weinheim, München: Juventus, pp. 544-568.

Schmidt, C. (2005): "Analyse von Leitfadeninterviews." In: Flick, U./von Kardorff, E./Steinke, I. (eds.): Qualitative Forschung. Ein Handbuch, Reinbek bei Hamburg: Rowohlt, pp. 447-456.

Schneidewind, U./Scheck, H. (2013): "Die Stadt als „Reallabor" für Systeminnovationen." In: Rückert-John, J. (ed.), Soziale Innovation und Nachhaltigkeit, Wiesbaden: VS, pp. 229-248.

Schroeter, K.R. (2005): "Pflege als figuratives Feld." In: Schroeter, K.R./Rosenthal, T. (eds.), Soziologie der Pflege: Grundlagen, Wissensbestände und Perspektiven, Weinheim: Juventa, pp. 85-106.

Schütte, J.D. (2012): "Soziale Inklusion und Exklusion: Norm, Zustandsbeschreibung und Handlungsoptionen." In: Huster, E.-U./Boeckh, J./Mogge-Grotjahn, H. (eds.), Handbuch Armut und soziale Ausgrenzung, Berlin: Springer, pp. 104-121.

Schütze, F. (1983): "Biographieforschung und narratives Interview." In: Neue Praxis 13/3, pp. 283-293.

Selke, S./Biniok, P. (2015): „Assistenzensembles in der Gesellschaft von morgen." In: VDE (ed.), 8. AAL-Kongress 2015, Berlin: VDE, pp. 50-57.

Staudinger, U.M./Schindler, I. (2002): "Produktivität und gesellschaftliche Partizipation im Alter." In: Schlag, B./Megel, K. (eds.), Mobilität und gesellschaftliche Partizipation im Alter, Stuttgart: Kohlhammer, pp. 64-86.

Strauss, A. (1993): Continual Permutations of Action, New York: de Gruyter.

Strauss, A./Corbin, J. (1990): "Grounded Theory Research: Procedures, Canons and Evaluative Criteria." In: Zeitschrift für Soziologie 19/6, pp. 418-427.

Tews, H.P. (1999): "Von der Pyramide zum Pilz. Demographische Veränderungen in der Gesellschaft." In: Niederfranke, A./Naegel, G./Frahm, E. (eds.), Funkkolleg Altern 1. Die vielen Gesichter des Alterns, Wiesbaden: VS, pp. 137-185.

Thomas, W.I./Thomas, D.S. (1928): The child in America: Behavior problems and programs, New York: Knopf.

VDE (2008): Intelligente Assistenz-Systeme – im Dienst für eine reife Gesellschaft, Frankfurt am Main: VDE.

Wahl, H.-W. (2002): "Lebensumwelten im Alter." In: Schlag, B./Mergel, K. (eds.), Mobilität und gesellschaftliche Partizipation im Alter, Stuttgart: Kohlhammer, pp. 48-63.

Wahl, H.-W./Schilling, O./Oswald, F. (2002): "Wohnen im Alter – spezielle Aspekte im ländlichen Raum." In: Walter, U./Altgeld, T. (eds.), Altern im ländlichen Raum: Ansätze für eine vorausschauende Alten- und Gesundheitspolitik, Frankfurt am Main: Campus, pp. 245-262.

Walker, A. (2002): "A strategy for active ageing." In: International Social Security Review 55/1, pp. 121-139.

Walter, U./Altgeld T. (2002): "Alter(n) im ländlichen Raum – Resümee und Ansätze für eine vorausschauende Gesundheits- und Altenpolitik." In: Walter, U./Altgeld, T. (eds.), Altern im ländlichen Raum: Ansätze für eine vorausschauende Alten- und Gesundheitspolitik, Frankfurt am Main: Campus, pp. 318-327.

Walter, U./Schneider, N. (2008): "Gesundheitsförderung und Prävention im Alter – Realität und professionelle Anforderung." In: Henson, G./Henson, P. (eds.), Gesundheitswesen und Sozialstaat, Wiesbaden: VS, pp. 287-299.

Wansing, G. (2005): Teilhabe an der Gesellschaft – Menschen mit Behinderung zwischen Inklusion und Exklusion, Wiesbaden: VS.

WHO (2002): Aktiv Altern: Rahmenbedingungen und Vorschläge für ein politisches Handeln, Wien: Bundesministerium für soziale Sicherheit, Generationen und Konsumentenschutz, Kompetenzzentrum für Senioren- und Bevölkerungspolitik.

Wichert, R./Norgall, T. (2009): "Natürlich altern ohne alt zu sein – Technische Lösungen für soziale Herausforderungen?". In: Zec, P. (ed.), Universal Design – Best Practice, Essen: red dot, pp. 56-63.

Wingenfield, K./Schaeffer, D. (2001): "Nutzerperspektive und Qualitätsentwicklung in der ambulanten Pflege." In: Zeitschrift für Gerontologie und Geriatrie 34, pp.140-146.

Ageing, Technology and Elderly Care: Assistive Technologies

Skripting Age – The Negotiation of Age and Aging in Ambient Assisted Living

CORDULA ENDTER

1. DESCRIBING THE FIELD: THE MISSING TELEPHONE

What does getting older mean in a postmodern society? Is it possible to stay autonomous, independent, self-determined? These questions underlie the discourse of Ambient Assisted Living (AAL) and ask if technological support – like intelligent wheelchairs or fall detectors, transponders or communication tools – is appropriate to assist older people.

Assistive technologies like AAL should enable persons to lead an independent life at home instead of in nursing institutions, to be mobile instead of confined in bed, open-minded instead of stubborn, socially and physically active instead of secluded. A new image of age and aging appears that is constructed socially, culturally, but also – and maybe foremost – politically. The young olds represent a new category of aging that is intrinsically linked to neoliberal and postmodern figurations of subjectivity, flexibility and autonomy (Dyk/Lessenich 2009) and can therefore be understood as an expansion of the neoliberal entrepreneurial self like it is discussed critically by Bröckling (2007). This perspective on AAL as a discourse of power in which age and aging are configured – following Steve Woolgar (1993) – or prescribed – according to Madelaine Akrich (1992) – is de-scripting: it uncovers a subjectivation of age and aging by promoting independence and autonomy through the use of assistive technologies.

I de-construct this black-boxed power regime in the following text by describing the practices of user integration in the design process of assistive technologies. By explaining the specific practices of user integration empirically, I aim at illustrating how age and aging are pre- and inscribed into the technical devices and thus pre-script age and aging as Akrich would argue

(Akrich 1992). To this aim, I want to give a first insight into the laboratory of an AAL project. But before I open the lab door, I want to situate my research context shortly: Ambient Assisted Living (AAL) refers mainly to smart and intelligent technologies that enable elderly persons to stay independent and self-determined in various areas of their life, such as health, housing, mobility, security and communication. Therefore, AAL technologies should be adaptive, usable, affordable, discreet and intuitive (Lindenberger 2007). Although smart technologies, especially for living and communication applications, are already quite common, the turn towards age and older users challenges the conventional engineering processes. Older users, especially if they are not as familiar with technology as younger generations, may present special physical, cognitive and emotional needs – for example, a reduced retentiveness or deficits in their motoric functions. These age-specific aspects challenge the development process; for example, menu navigation has to be kept simple and non-hierarchical to preserve the memory capacity of the user. Consequently, the development process itself gets more complex, takes more time and requires knowledge from non-technical fields like gerontology, psychology or geriatrics. In 2008 the German Federal Ministry of Education and Research (BMBF) reacted to demands of engineers, computer scientists, companies and future users and set up its funding program "Age-based Assistive Systems for Living Healthy and Independent – AAL"[1] with its first announcement "Living Self-Determined" (BMBF 2008)[2]. Numerous announcements have already been published (MTIDW 2015) that have broadened the scope of AAL-technologies. When I started to become interested in this topic in 2013, the first funding period had ended. A lot of high-tech cutting-edge technologies had been invented, but most of them failed when entering the market (Marschollek/Künemund 2014). Why? Following evaluations, several reasons – time, funding and knowledge – became clearer and it turned out that the nescience about the user and his or her everyday

---

1 | The initiative of the BMBF was one of the main starting points for the Federal Government's research agenda "The New Future of Old Age", which was announced in 2011 with the "aim to conduct research that will encourage the development of new solutions, products, and services to improve the quality of life and social participation of older people. Discovering the hidden treasures of an ageing society will benefit all generations for demographic change" (BMBF 2011a).

2 | To the specifics of AAL in Germany see also Künemund/Tanschuss (2013).

life was the biggest challenge to tackle. The project executing organization, VDI-VDE-IT, reacted to this deficit and revised its announcement: User-integration – carried out through user-centered design[3] – became an integral part of the funding strategy.

This turn towards the user was also my starting point for undertaking ethnographical research on AAL. Therefore, I accompany different AAL projects by doing participant observation in the projects, interviewing staff and test persons and analyzing the project documents systematically. AAL conferences and workshops complemented my field of studies. Additionally, I analyzed the funding announcements, programs and publications of the BMBF and VDI-VDE-IT. In this article, I will focus on the integration of users in AAL projects by describing in more detail the implementation of user-centered design in one project.[4] The leading questions are therefore: firstly, how does an older person become a test user? And secondly, to what extent can user-centered design be put into practice? To answer these questions I would like to start by giving a short impression about the user-testing situation:

3 | The integration of users in the design process can be carried out by means of different methodological approaches like, for example, user-centered design, participative design or value-sensitive design. Whereby participative design and value-sensitive design are common approaches in Scandinavia or the Netherlands, German usability studies prefer the user-centered design approach. It is a certified procedure (DIN ISO 9241-210) to ensure the usability and usefulness of a technical system or device by considering needs and requirements of potential users. Therefore, its usability, operationalized by effectiveness, efficiency and satisfaction, is measured in specific periods – the so called formative and summative evaluation – throughout the design process. Here, test persons simulate the behavior of potential real users by solving tasks on a prototype in a laboratory setting and evaluate the above mentioned criteria by answering a questionnaire. In the formative evaluation, the results of the usability tests will be considered in the next steps of the design process. The summative evaluation finalizes the design process and its results are immutable.

4 | The following description of the project, its laboratory setting, and the involved persons like Mr. Schreiner or Mr. Wolfe, are based on my ongoing field research in AAL mentioned above. The name of the project and the persons quoted in this text were anonymized by the author. The participant observation and the informal talks with the participants and project staff took place between March 2013 and January 2014.

"At home, I would simply call", Mr. Wolfe says, "I would simply pick up the phone and dial the number and then I know if one is there or not." But there is no telephone on the table in the laboratory. Thus, Mr. Wolfe tries again to activate the screen of the flat device lying in front of him on the table. "Push the button and then pull the arrow to the left softly", he whispers, while he is pushing the white button again to pull the arrow to the left. But again, nothing happens. Mr. Wolfe moves nervously on his seat, puts the chair a bit closer to the table, dries his hands on his trousers and tries a third time. It works. "At last!" Mr. Wolfe breathes a sigh of relief. The arrow disappears and a new screen opens. "That's why, the CASEtab is so fantastic. You just take it on and – let's go!" Mr. Schreiner comments on the scenery. Mr. Schreiner is the usability expert who invited Mr. Wolfe some time ago to test a new communication tool that should enable elderly persons without any computer knowledge to communicate via an intuitive and age-based web2.0-platform that is easy to use. I got to know both men during my participant observation in the usability laboratory. In total there were 45 people who tested the platform and evaluated its usability and appropriateness for implementation in the users' everyday lives. Therefore, the users had to visit the laboratory once for approximately two hours. In this time, they solved a set of 16 tasks including writing an e-mail, making an appointment, inviting a friend for a walk, checking out new theatre programs and changing dates on their profile. All the tasks had to be carried out on a tablet PC which was connected to the Internet. An application had been installed on the tablet, which included a reminder for medication, appointments or other events. Besides, it featured a map and a list with local shops like banks, grocery stores, pharmacies or ambulances that was linked to the map. It also included a floor plan of the user´s flat with household devices and furniture like lamps, cooker, windows and the door. All these objects were equipped with sensors that enable the user of the app to control the functions of the objects via the application. For example, if Mr. Wolfe is sitting in his chair and somebody is ringing the doorbell, he doesn't have to get up: he just opens the app, opens the program, pushes the icon with the door and then decides to open it or not while seeing the face of the visitor on his screen.

The CASEtab was designed for older users with less experience in computer usage in an iterative process where users were addressed to take part in the design process (UCD). Therefore, older people were invited to take part in different stages of the design process. First, a group of twenty test users was asked for their needs in terms of technical support in their activities of daily life; afterwards, another group was asked about their demands. Based on the interview material, the iterative process of prototyping was started without any user participation. The following time, a group of five users participated in the evaluation of the first prototype (formative evaluation) during the design process. The last time users were invited was the end-evaluation, where I took part and met Mr. Wolfe and Mr. Schreiner.

Mr. Schreiner is conducting several AAL projects, "but it is still a challenge to find participants that are motivated to take part in several sessions, answering several questions and doing several tests", he summarizes. Mr. Wolfe retired two years before the project and is living near the laboratory. He already took part in other tests addressing cognitive or motoric competences, but he never used a tablet before. The laboratory is situated on a research campus of a German university where a lot of user tests – not only age-specific ones – are carried out. The majority of the recruited test users is registered in a database of the research center and this data is used for experimental settings, questionnaires and tests. My presence was rarely perceived as unfamiliar, although I introduced myself as a cultural anthropologist doing empirical research for my PhD.

The end-evaluation is typically the last test-situation of the prototype and functions to confirm its usability. This means its efficiency (tasks correctly solved), its effectiveness (time to solve the task) and the satisfaction of the users are quantified to decide whether the design process ended successfully or not. Success in this case means that the prototype is considered to be competitive on the market. Hence, the end-evaluation is a fragile and unstable stage in the design process, where not only the device itself is evaluated but also the work and thus the knowledge, the creativity and the abilities of the project team. If the prototype fails, corrections are no longer possible, since funding is running out. Consequently, the project, or rather, the socio-technical ensemble following Actor-Network Theory (ANT), is extremely careful to avoid any kind of failure. Therefore, the project staff works on stabilizing the prototype technically as well as socially by controlling the evaluation and configuring the role of the test users in the test scenario. How do they carry out this invisible work of controlling? Obviously, they try to minimize technical bugs before or even during the tests: therefore, Mr. Schreiner supervises the test and intervenes if something unexpected happens, such as a breakdown of the application, long loading time or a mistake in the menu-driven operation. In addition, they try to modify the test situation by designing tasks that are solvable for the majority of the users or by giving them enough time to solve them. These practices are not as obvious as checking for technical or operational bugs. Instead, these are non-transparent for the users, who are and should be unaware of the designed context of the test situation, and invisible in the test manuals or protocols. It is the task of the supervisor to make these practices invisible.

Thus, the test supervisors are powerful actors in the socio-technical ensemble, not only because they design the technical object, but also because they are able to black-box the difficulties and hindrances. Test users at this stage challenge the success of the evaluation because their testing behavior, their critique and their recommendations are not only a re-instabilization of the prototype, but also because their results are a materialization of failure. Hence, user tests are a discrete practice of control, configuration and pre-scription. The construction of age-specific technologies is thus a social practice in which the final end-product is the result of diverse negotiations by human as well as non-human actors. *Doing age* by negotiating age-specific technologies means in this context *doing age by designing age-specific technologies* that inscribe age and aging into the technology by, first, determining age-specific technologies as age-specific and, second, addressing their users as aged. To describe the various interactions and arrangements that are needed to design AAL and thus to construct a socio-technical ensemble, my approach was to follow Mr. Wolfe and Mr. Schreiner in their negotiations.

2. PRE-SCRIBING AGE: THE DISCOURSE OF AMBIENT ASSISTED LIVING

Over the last two decades, Western societies like Germany seem to have transformed into technological societies. Computer-based technologies in particular have entered the everyday life of people, regardless of social class or cultural habits. At the same time, these societies have had to face tremendous demographic changes. On the one hand, the number of people older than 65 in Germany is statistically rising, while on the other hand the senior population itself is getting older. This growing sector of older people is required to stay active, engaged and informed to gain societal participation and recognition. At this point, technology is introduced as an instrument of empowerment. By using smart assistive devices, older people should engage in society and remain autonomous and mobile while staying in their familiar environment in terms of living and housing. Consequently, AAL technologies represent a strategy to maintain this activity potential and broaden it towards new fields of application like health, communication and information or security. Thus, AAL is not only empowering but also normative and hegemonial.

Already in the early 2000s, the sociologist Stephen Katz argued from a critical gerontologist perspective that the association of activity with well-being in old age became "so obvious and indisputable that questioning it within gerontological circles would be considered unprofessional, if not heretical" (Katz 2000: 136). Katz points out that the idealization of activity in gerontological discourse helped to establish neoliberal regimes of activity that "manage everyday life in old age" (ibid: 142) in terms of physical and cognitive activity. He states that "[m]ost gerontological and policy discourses pose activity as the 'positive' against which the 'negative' forces of dependency, illness, and loneliness are arrayed" (ibid: 145). The activation of old age subjectifies old age and at the same time allies "their active subjective efforts at maintaining autonomy and health with the wider political assault on the risks of dependency" (ibid: 146; Dyk et al. 2010; Denninger/Lessenich 2012; Schroeter 2000). Following Bruno Latour's argument "that we cannot understand how societies work without understanding of how technologies shape our everyday lives" (Latour 1992: 151), critical aging studies have to ask how assistive technologies are adjusted to the everyday lives of older people, why they are adjusted, and in which ways.[5] De-constructing the co-construction of aging and technology is necessary to de-mask AAL as a neoliberal strategy of subjectifying risk and dependency to old age instead of reminding welfare policies of their duty.

AAL technologies, as technologies of everyday life, are supposed to remind their users to take their pills at the right time, control their household devices, monitor their bio-medical parameters or organize their daily activities. Although it might be useful for everyone to have a control system of his or her daily activities, this seems to be especially useful for older people. AAL is marketed as age-specific, while at the same time age is equated with being in need of technological assistance. Thereby, it is the relatedness to age that distinguishes AAL technologies from convenient smart household de-

---

5 | Although critical cultural gerontology scholars like Stephen Katz (2000) or foucauldian aging studies scholars like Simon Biggs and Jason L. Powell (2001) do not specifically work on age-based technologies, their critical approach seems to be very productive for a critical view on AAL.

vices, but it is also the development of smart devices suiting the needs of the elderly that produces these needs in the first place. Aside from the notion that this premise is already creating stereotypes, it is also powerful by prescribing age and aging into scripts and programs of AAL devices. This inscriptional work materializes itself again in displays, touch pads or transponders. Different stakeholders – government agencies, scientists, entrepreneurs, care providers – define age and aging as a demographic challenge to claim legitimate involvement in this issue. Here, the discourse of demographic change that is often articulated as a threat or burden (e.g. Butterwegge 2006; Grebe 2012) seems to be quite attractive to promote technological innovation for the elderly, but the real objective actually seems to be promoting economic development. Seen in this light, AAL technologies are less a demographic tool-kit mastering the challenges of demographic aging than basically a federal promotion of national economy.

Therefore, potential users have to be addressed and actively integrated in the design process. This is, however, a challenge, as Mr. Schreiner already claimed, while his colleague Mr. Miller points out: "It's nothing else than prose."

3. Inscribing Age: How Users Matter

Although over the last two decades the motto that 'users matter' has become evident in a number of different areas of technology studies (e.g. Oudshoorn/ Pinch 2008), aging is a *terra incognita* for most computer scientists and engineers. Consequently, it seems to be useful to become acquainted with it. The question is how. Within informatics and ergonomics, usability tests take over the role of "contact zones", where technical devices, technicians and future users get in contact with each other by testing and evaluating the innovated products. In AAL, this contact zone is produced by user-centered design, whereas the term user-centered is ambivalent. Although it proposes that users play a key role in the innovation process, it is not certain if they fulfill this role actively as participants or passively as part of a preconceived idea of the users' issues, demands and needs. During my fieldwork I started from the premise that user participation can be both: an active participation of users, but also an idea in the users´ minds. The users' agency depends on the situation and the setting of the design process: if the projects are funded by the Federal Ministry of Education and Research (BMBF), they are obliged to work user-centered (BMBF 2011b). Due to different reasons, which I can enumerate here only briefly, this obligation is carried out in very different ways.

Although "the old view of users as passive consumers of technology has largely been replaced and along with it the linear model of technological innovation and diffusion" (Oudshoorn/Pinch 2008: 543), the idea of making users central to innovation is driven, first of all, by economic reasons, and not because it is reasonable to integrate users in the innovation process. On the contrary, user-centered design can cause additional costs, but it increases the probability that the product will be successful on the market. Therefore, user-centered design was introduced as a promise for a better understanding of users' needs and demands and translating them into technical features, which should raise the commercial attractiveness of the final product.

When taking a look at the production of UCD, the ambivalence remains or even gets more complex, because it is in most cases delegated to actors who do not possess the competence for producing user-centered design. Not all project members are familiar with usability trials or qualitative methods from social sciences like interviews, questionnaires or observation. Although the BMBF has already reacted on this deficit by announcing that project members have to be interdisciplinary and at least one position has to be full filled by a social scientist, this scientist is not automatically the one responsible for UCD. Instead, the UCD test is mostly conducted by computer engineers, psychologists or other project members. Hence, the problem of user integration still remains and gets even more problematic in view of the heterogeneous targeted group.

Thinking of age and technology as together is still somehow paradoxical: older people currently do not have a very wide biographical experience with information and communication technology because technology did not play a big role in their work experience or in their leisure activities (e.g. Czaja et al. 2006; Czaja/Lee 2007; Mollenkopf 2008). Although this imbalance will change within future cohorts, it is a challenge for the current users and that is why their participation is crucial for developing technologies, which should not only fit their needs but also be used by them properly. Therefore, the test-users need to be representative of the actual users, and this raises the question of selection (e.g. Collins/Evans 2002). Who should be included by excluding whom? Who is representing whom? The selection process often is not objective: instead, it follows practical aspects like experience, contact possibilities and time. The test users for the CASEtab were selected by using a database of people interested in performing the test, who received an invitation to the institute laboratory. Most of the test volunteers were living nearby, had retired and were members of a so-called "Senior Academy", so they were familiar with testing situations and in some cases even with the laboratory. Their motivation for taking part in the test was often not an interest in the age-specific CASEtab, but rather curiosity about getting tested or just time availability. Although this may sound trivial, it manifests that those who took part in the test were mobile, informed and healthy enough, as well as with enough free time and self-esteem to try out an unfamiliar testing situation. All the other members of this age group remain invisible not only in the test itself, but also in the final report, in articles or presentations and in the innovation process – and, therefore, in the finished product. Which users are inscribed when the majority of potential users in need of the product are not visible in the innovation process? Or more specifically, which ideas of users are inscribed?

User integration is a black-boxed process. While the official report includes short descriptions of the sample by naming variables like age, economic status and professional background, it is not reported whether the test persons were already familiar with test situations or whether they had to come to a laboratory even when the technical device they tested is a tablet and, therefore, mobile. Black-boxing the selection of test persons and hence the conditions of the usability evaluation is congruent with the strategy of black-boxing the outcomes of the project. Of course, efficiency, effectiveness and satisfaction of CASEtab are manifested in tables, graphs and numbers and translated in a business model, but the question is still whom they target.

The test users of the CASEtab were on average active, physically and cognitively healthy, and received a satisfying pension, but are they representative of those in need for such a technology? CASEtab should assist users who are unfamiliar with information technologies and communication techniques like chat, email, video-calling or Internet by providing a smart and user-friendly device which foregoes complex menus, difficult commands or detailed displays. Instead, the display components are repetitive, simplistic and clearly structured.

Although user-centered design can be understood as a trial to control users as well as their usage, there is still some potential for resistance and a certain amount of uncertainty that remains: *"Nothing in a given scene* can prevent the inscribed user or reader from behaving differently from what was expected. [...] There might be an enormous gap between the prescribed user and the user-in-the-flesh" (Latour 1992: 161, emphasis in original). Black-boxing this gap, the engineers of CASEtab decided to invite potential users to the laboratory in order to test the platform instead of observing their daily routines of household duties, shopping or communication in their familiar surroundings. Physically confronting real users, researchers were confronted with the gap between their imaginations and their inscriptions of imagined users and the real users – or, as Latour names them, the "users-in-the-flesh" (Latour 1992: 161).

Predictably, the user tests increased that gap. The practical use of the tab revealed a lot of incorrect or overlooked inscriptions of age-specific usage. CASEtab provides a lot of digital features that are designed in comparison to their analog originals, such as an address book. The contacts are listed alphabetically and a frame around each contact separates it from the next: each contact is defined with a photo, name and surname, email address and telephone number. Upon touching a contact, another menu opens where they find more information about the person – for example date of birth or address. At the same time, they can touch the telephone button to call the person directly or the envelope button to write this person an email – this also confused the users, who expected a writing program where they could write down a letter and not a messenger formula. The biggest problem, however, was not the display of information or the retrieval of a contact, but moving through the display. The users in the laboratory needed either several minutes or the help of the test supervisor because they were not familiar with the gesture of scrolling that enabled them to move the display and simultaneously their contacts. Although the gesture of scrolling was explained shortly at the beginning and the test supervisor provided hints that should remind them of it, only a small part of the 45 test users remembered this gesture and applied it correctly and efficiently. In some cases, it happened by chance that the users revealed that the display can be moved by touching it but then forgot to do the same in the next task. Hence, instead of solving the task, users got frustrated and unsure of their competences. Some of them tried to touch all the available buttons, photos or icons, thinking that this might set off an action, but this strategy resulted in a loss of orientation. Mapping them deeper and deeper in the menu without them knowing how to do it and why made them frustrated, anxious and helpless. If they interrupted the task and the test supervisor explained the scrolling function to them, they became even more confused because it was such a simple gesture and they could not master it.

Here the inscription failed, as the aged users could not solve the task: "It is only when the script set out by the designer is acted out – whether in conformity with the intentions of the designer or not – that an integrated network of technical objects and (human and nonhuman) actors is stabilized" (Akrich 1992: 222). The script of the CASEtab was not carried out by the users. They could not operate the display because, on the one hand, the designers were not informed or did not realize that the simple task of scrolling up and down to move the display and make information visible is unfamiliar to those users. On the other hand, the technical device was not programmed to provide hints. Neither was it attached with a slider bar on the left side, nor did the display switch up and down when it was opened or when the arrow jumped up and down. The interaction failed and the network of human and nonhuman actors remained unstable because the users were not able to read the script and thus use the tab. Although this missing feature can be added easily by designing visual aids in the display, this scenario reveals the complexity of designing human-computer interaction as well as the impossibility to fully inscribe age-specific usage. Consequently, both the human-computer interaction and the sociotechnical ensemble cannot be stabilized in the usability trial. Thus, the idea of smart technical assistance remains unstable as well. What is then questioned is the construction of competence.

In this context, competence is a precondition to use the developed device adequately, but competence is also an "articulation" (Moser/Law 2003: 491) of the aging subject or, as Akrich and Latour describe it, a "pre-scription" (Latour 1992: 178) of the aging actor: "We call *pre-inscription* all the work that has to be done upstream of the scene and all the things assimilated by an actor (human or nonhuman) before coming to the scene as a user or an author" (ibid, emphasis in original). Pre-inscriptions of competence are produced in the discourse of active aging by assuming that it is a natural need of older people to work actively and voluntarily on their abilities and competences, as Paul Baltes and colleagues suggest in their model of selection, optimization and compensation (SOK-model), where the aging subject is doing selection, optimization and compensation work to balance their loss of abilities by focusing on a manageable and necessary repertoire of abilities (Baltes/Baltes 1990). The fact that this repertoire is socially constructed by the activation discourse is as concealed as the conditions of this 'competence work'. How "'the competent and abled person' is constructed (or not) under specific circumstances and how it is that he or she is constructed (or not) in relation to new media technologies" (Moser/Law 2003: 491) is the question that matters when discussing the "articulation work" (Fujimura 1987: 260) of AAL and its concealed hegemony. It is the invisible work of pre-inscription that enables engineers to "bet on this predetermination when they draw up their prescriptions" (Latour 1992: 178). In her work of *de-scription,* Akrich shows that "the ease with which the actants assumed in the design of the object are related to those that exist in practice is partly a function of decisions by designers" (Akrich 1992: 207). Therefore, de-scribing user-centered design as *taking* decisions through design and thus determine what is an appropriate usage (or not) or what is a competent user (or not) – and ultimately, what is successful aging (or not) – is the necessary scientific task of giving a voice to those who are voiceless (e.g. Moser/Law 2003: 494). This of course means to focus on the technical device as an actor to whom competence is delegated or rather inscribed.

In the case of CASEtab, age is inscribed in terms of information, communication and control. The underlying image is that aged users unfamiliar with smart information systems like tablets or smartphones are ambitious to use them to stay informed and engaged through new media technologies. Therefore, the designers invent an application that reminds them, connects them, informs them, and thus materialize their aging subjectivity in an age-specific device to assist their everyday activities technically. Using CASEtab first of all articulates its users as old and in need for smart technical help. At the same time, it reinforces these imaginations of age and technology by deciding which functions CASEtab should fulfill or not. To conceal this decision-making process is a powerful strategy to produce asymmetries by announcing the opposite (e.g. Garrety/Badham 2004). "Why shouldn't I just call him?", Mr. Wolfe is asking after several misleading trials to activate the video telephony. Mr. Schreiner answers: "Because it's so much easier!" Mr. Wolfe shakes his head. "Why should this be easier", he asks himself in amazement, "it would be so much faster to pick up the phone". After a while, he completes his thought: "It would make me feel nervous, only because if I want to make a telephone call, I would have to shave, brush my hair and put on a shirt. Why should somebody see me like this?" Mr. Wolfe refuses to take over the explanation of Mr. Schreiner and hence refuses the inscriptions made by the designers: he articulates his autonomy in terms of skepticism and refusal (e.g. Moser/Law 2003). In this case, the misleading concept of UCD becomes obvious. Mr. Wolfe's reaction on the idea of video telephony that he expresses clearly towards Mr. Schreiner has to be rejected by Mr. Schreiner because the prototype is already finalized and further modification is not planned or funded. Hence, Mr. Schreiner has to convince Mr. Wolfe that video telephony is "so much easier" to stabilize the innovation work that is materialized in CASEtab. Hence, he is doing articulation work by concealing refusal or critique and attempting reassurance. He points out that the effort of calling by using the application is faster, less complex and more personal, but by doing this he misses Mr. Wolfe's point that the inscription of communication behavior mismatches his everyday practices of communication.

The *ageskript,* as I call the inscription of age into technical devices like AAL, fails because the evaluation of the prototype is neither participative nor symmetrical. "Only by describing both the production task and the hidden task in articulation, together and recursively, can we come up with good analysis of why some systems work and others do not." (Star 1999: 387) Following the symmetrical paradigm of de-scribing user-centered design brings all relevant actors with their practices into focus. Here, ethnography can play an important role by observing practices, performances and interactions of human and nonhuman actors in the field. Following each actor in his multiple sites – like George Marcus' approach of multi-sited ethnography suggests – means to trace "things in and through contexts" (Marcus 1995: 107). Thus, fieldwork makes "the invisible matters of causes, the regimes, the blank spaces, demarcations and hierarchies visible" (Windmüller/Binder/Hengartner 2009: 16). By moving in and through these different sites, relations and articulations of subjectivity of AAL can be made visible, since "[s]till, no scene is prepared without a preconceived idea of what sort of actors will come to occupy the prescribed positions" (Latour 1992: 161). Doing participant observation is therefore an appropriate method to de-script the hidden knowledge regimes, the translations and delegations of agency in this hybrid constellation of human-computer interaction. Being a visible observer in an invisible field makes the invisible visible by "valorizing previously neglected people and things" (Star 1999: 379).

4. Circumscribing Age: Black-Boxing Competence

As I argued in a previous section, the interaction of project members (engineers, computer scientists, designers, etc.), test users and technical devices has to be de-scripted as *doing age by technology.* Therefore, re-constructing the production of an AAL device by describing the inscribed power regimes and articulations is conducive to understanding how AAL technologies configure age and aging.

But not only the interaction between designers/computer scientists/engineers and test users in the user-centered design process is multi-sited and transitive; it is also marked by uncertainty, nescience and, in some cases, ignorance, so the project work itself becomes asymmetrical. I would like to describe this asymmetry by means of an example.

The psychologist involved in the production of CASEtab claimed that a special colored background is useful for discriminating objects on the screen: she could verify her claim empirically by referring to scientific studies. The designer involved in the production of CASEtab objected that the suggested colors would not be attractive. The software engineer involved in the production of CASEtab was not willing to change the color again. What happened is that the psychologist repeated her arguments in every meeting, made the empirical studies accessible for all project members and suggested design alternatives. She made all her arguments visible by loading them up to the company server, where everyone involved in the production of CASEtab had access, but nobody opened that folder except the psychologist herself.

In the next telephone conference, the psychologist brought up the unsolved question of color. The other team members and project partners expressed their conformity immediately by shouting: "Yes, that's a good point", or "It's good that you think about it", or "We should keep it in mind". Then, they continued with the next topic. The psychologist felt frustrated and powerless. She was only one member representing one partner in a constellation of seven, whereby the other six ones were technical partners. At the end, the software engineer finalized the color by inscribing it into the technical features of the application: he felt legitimated to do so via his competence and his position in the project as an executive actor.

His strategy of keeping the design process manageable by deciding pragmatically is a very common strategy: other strategies are indirect communication, deceleration of decision-making, and refusing designation. These indirect communication strategies foster the exclusiveness of knowledge and expertise. It is not only necessary to question how engineers delegate power to AAL technologies; it is also necessary to de-scribe the power relations in the projects to understand the inscription of age. In relation to the above discussed question "how it is that 'the competent and abled person' is constructed (or not)" (Moser/Law 2003: 491), competence again gains relevance. It is through the interaction of the psychologist, the software engineer and the device in the working context of the project where competence is used to mark the boundaries of agency and power. Although the psychologist is – due to her scientific background – assumed to be the expert, it is the software engineer who finally decides which color to take. To legitimate his decision, he welcomes the work of the psychologist cursorily and at the same time ignores her arguments and work samples. It is his executive function in the project setting that enables him to do so. No other project partner is able to write the software program: this exclusiveness is a powerful means to prioritize his argument and to black-box the others.

Here it must be questioned to what extent the role of nontechnical project partners can be compared to those of test users. The described interaction indicates a hierarchical imbalance between the project partners that has not occurred by chance, but is rather produced actively to enforce the innovation process in the designated direction. This is an assumption that is neither developed in the design process (by integrating test users) nor communicated in the project meetings (through equal cooperation and transparent decision making), but it is obvious to the software engineer because he translates the concept of CASEtab into the technical device by writing the script and therefore delegating agency to the object. This mediating work is deeply bounded to the technical options available to the engineer (e.g. Akrich 1992), the scripting knowledge he embodies and at least his motivation to be innovative. These circumstances are not communicated in the project: rather, it is the work of the engineer to make them invisible and the work of the ethnographer to de-script them and make it visible that "[t]he obduracy or plasticity of objects, […], is a function of the distribution of competences assumed when an object is conceived and designed" (ibid: 207).

5. SKRIPTING AGE: NEGOTIATIONS AND OSCILLATIONS

What do these observations mean for a scientific approach of understanding AAL as *doing age by technology*? How can the scripting of age be de-scribed from a cultural aging studies perspective? And how can the normative interplay of power, innovation and age become visible in the research? I argued that the multi-sited (Marcus 1995) interactions of the sociotechnical ensemble (Latour 1992) have to be de-scripted (Akrich 1992) in order to re-construct the black-boxed processes that enabled the project partners "to turn technical objects into black boxes" (Akrich 1992: 221). It is not before they are black-boxed that they can be stabilized and therefore "become instruments of knowledge" (ibid). The translation of complexity into an easy-to-use technical device like CASEtab can only be carried out through the interaction of different human and nonhuman actors. Here, I introduced user integration by user-centered design as a possibility to de-script how "technical objects and people are brought into being in a process of reciprocal definition in which objects are defined by subjects and subjects by objects" (ibid: 222). In the context of AAL, this means to de-scribe how the subjectivity of aged users is inscribed into AAL and, at the same time, how AAL determines the subjectivity of aged users through the inscriptions. I argued that the integration of users to stabilize the translation is a controlled process in which the usability tests produce a contact zone where the different imaginations of AAL are negotiated; however, I also argued that this process of negotiating age and aging is controlled by the project team that is conducting the tests. This control is necessary because the real users are not conforming to the imagined users that underlie the innovation process: these ideas about users are non-empirical, non-theoretical and non-scientifically based. They are assumptions that have materialized in the technical devices: therefore, their authentication in the evaluation process is not possible.

The tests are thus doomed to failure, but failure is not acceptable and hence different strategies – controlling the test, designing solvable tasks, black-boxing critique or refusal – are applied to guarantee success. This is due to the false idea that AAL technologies are easy and quick, innovative and profitable objects that automatically strengthen older people's autonomy and societal engagement when they are used by them. Instead, the innovation of assistive technologies should be understood as an oscillatory process in which the prototype is a materialization not only of ideas and imaginations of age and aging, but also of social practice. AAL is something that is produced interactively by nonhuman actors – like the technical devices themselves, but also the technological infrastructure in which the innovation process is embedded – and by human actors like, for example, Mr. Wolfe or Mr. Schreiner. Their interaction is multi-directional, iterative and complex and takes time, but it is at the same time necessary for developing cutting-edge technologies. Therefore, user integration is inclusive.

Taking it seriously demands making these interactions visible: this approach would bring about coherence, reflexivity and transparency back into the design process. Furthermore, it would bring participation into the laboratory and destabilize the normative power hierarchies of AAL. Then designing AAL could become not only a network of innovation or an innovative object for older users and market interests; it could also be an experimental space to think of age and aging differently.

REFERENCES

Akrich, M. (1992): "The De-Scription of Technical Objects." In: Bijker, W.E./Law, J. (eds.), Shaping Technology/Building Society: Studies in Sociotechnical Change, Cambridge: MIT Press, pp. 205-225.

Baltes, P.B./Baltes, M.M. (1990): "Psychological Perspectives on Successful Aging: The model of selective optimization with compensation." In: Baltes, P.B./Baltes, M. (eds.), Successful Aging: Perspectives from the Behavioral Sciences, Cambridge: MIT Press, pp. 1-34.

Biggs, S./Powell, J.L. (2001): "A Foucauldian Analysis of Old Age and the Power of Social Welfare", In: Journal of Aging & Social Policy 12/2, pp 93-112.

Bröckling, U. (2007): Das unternehmerische Selbst, Frankfurt am Main: Suhrkamp.

Butterwegge, C. (2006): "Demographie als Ideologie? Zur Diskussion über Bevölkerungs- und Sozialpolitik in Deutschland." In: Berger, P.A./Kahlert, H. (eds.), Der demographische Wandel. Chancen für die Neuordnung der Geschlechterverhältnisse, Frankfurt am Main/New York: Campus, pp. 53-81.

Czaja, S.J./Charness, N./Fisk, A.D./Hertzog, C./Nair, S./Rogers, W.A./Sharit, J.(2006): "Factors Predicting the Use of Technology: Findings From the Center for Research and Education on Aging and Technology Enhancement (CREATE)." In: Psychology and Aging 21/2, pp. 333-352.

Czaja, S.J./Lee, C.C. (2007): "The impact of Aging on Access to Technology." In: Universal Access in the Information Society 5/4, pp. 341-349.

Collins, H./Evans, R. (2002): "The Third Wave of Science Studies. Studies of Expertise and Experience." In: The Social Studies of Science 32/2, pp. 235-296.

Denninger, T./Lessenich, S. (2012): "Rechtfertigungsordnungen des Alter(n)s." In: Soziale Welt 63, pp. 299-315.

DIN ISO 9241-210 (2010): Prozess zur Gestaltung gebrauchstauglicher interaktiver Systeme. Berlin.

Dyk, S.v./Lessenich, S. (eds.) (2009): Die jungen Alten. Analysen einer neuen Sozialfigur. Frankfurt am Main: Campus.

Dyk, S.v./Lessenich, S./Denninger, T./Richter, A. (2010): "Die ‚Aufwertung' des Alters. Eine gesellschaftliche Farce." In: Mittelweg 36/5, pp. 15-33.

Fujimura, J. (1987): "Constructing 'Do-able' Problems in Cancer Research: Articulating Alignment." In: Social Studies of Science 17, pp. 257-293.

Garrety, K./Badham, R. (2004): "User-Centered Design and the Normative Politics of Technology." In: Science Technology Human Values 29/2, pp. 191-212.

German Federal Ministry of Education and Research (BMBF) (2008): Living Self-Determined, http://www.bmbf.de/foerderungen/12394.php (last access: 10.02.2015).

German Federal Ministry of Education and Research (BMBF) (2011a): The New Future of Old Age, http://www.das-alter-hat-zukunft.de/en (last access: 10.02.2015).

German Federal Ministry of Education and Research (BMBF) (2011b): Mobility in old age, http://www.bmbf.de/foerderungen/15268.php (last access: 10.02.2015).

Grebe, H. (2012): "'Über der gewonnen Zeit hängt eine Bedrohung.' Zur medialen Thematisierung von hohem Alter und Demenz: Inhalte, Strukturen, diskursive Grundlagen." In: Kruse, A./Rentsch, T./Zimmermann, H.-P. (eds.), Gutes Leben im hohen Alter. Das Altern in seinen Entwicklungsmöglichkeiten und Entwicklungsgrenzen verstehen, Heidelberg: Akademische Verlagsgesellschaft, pp. 97-108.

Katz, S. (2000): "Busy Bodies: Activity, Aging, and the Management of Everyday Life." In: Journal of Aging Studies 14/2, pp 135-153.

Künemund, H./Tanschus, N.M. (2013): "Gero-technology: Old age in the electronic jungle." In: Komp, K./Aartsen, M. (eds.), Old Age in Europe: A Textbook of Gerontology, New York: Springer, pp. 97-112.

Latour, B. (1992): "Where are the Missing Masses? The Sociology of a Few Mundane Artifacts." In: Bijker, W.E./Law, J. (eds.), Shaping Technology/ Building Society: Studies in Sociotechnical Change, Cambridge: MIT Press, pp. 151-180.

Lindenberger, U. (2007): "Technologie im Alter: Chancen aus Sicht der Verhaltenswissenschaften." In: Gruss, P. (ed.), Die Zukunft des Alterns. Die Antwort der Wissenschaft, München: C.H. Beck, pp. 220-239.

Marcus, George E. (1995): "Ethnography in/of the World System: The Emergence of Multi-sited Ethnography." In: Annual Review of Anthropology 24, pp. 95-117.

Marschollek, M./Künemund, H. (2014): „Gerontechnologie zwischen Akzeptanz und Evidenz." In: Zeitschrift für Gerontologie und Geriatrie 47, pp. 639-640.

Mensch-Technik-Interaktion im demografischen Wandel (MTIDW) (2015): Overview Announcements, http://www.mtidw.de/ueberblick-bekanntmachungen (last access: 10.02.2015).

Mollenkopf, H. (2008): "Neue technische Entwicklung und Erhalt der Selbständigkeit im Alter." In: Kuhlmey, A./Schaeffer, D. (eds.), Alter, Gesundheit und Krankheit, Bern: Huber, pp. 225-244.

Moser, I./Law, J. (2003): "'Making Voices': New Media Technologies, Disabilities, and Articulation." In: Liestol, G./Morrison, A./Rasmussen, T. (eds.), Digital Media Revisited: Theoretical and Conceptual Innovation in Digital Domains, Cambridge: MIT Press, pp. 491–520.

Oudshoorn, N./Pinch, T. (2008): User-Technology Relationships: Some Recent Developments, http://csid.unt.edu/research/trw/User-Technology%20Relationships,%20Some%20Recent%20Developments.pdf (last access: 30.07.2014).

Schroeter, K.R. (2000): "Die Lebenslagen älterer Menschen im Spannungsfeld ‚später Freiheit' und ‚sozialer Disziplinierung': forschungsleitende Fragestellungen." In:. Backes, G.M/Clemens, W. (eds.), Lebenslagen im Alter. Gesellschaftliche Bedingungen und Grenzen, Opladen: Springer, pp. 31-52.

Star, S.L. (1999): "The Ethnography of Infrastructure." In: American Behavioral Scientist 43, pp. 377-391.

Windmüller, S./Binder, B./Hengartner, T. (eds.) (2009): Kultur – Forschung. Zum Profil einer volkskundlichen Kulturwissenschaft. Berlin: Lit.

Woolgar, S. (1991): "Configuring the User: The Case of Usability Trials." In: Law, J. (eds.), A Sociology of Monsters: Essays on Power, Technology and Domination, London: Routledge, pp 58-100.

Making Space for Ageing: Embedding Social and Psychological Needs of Older People into Smart Home Technology

BARRY GUIHEN

1. AGEING AT HOME IN EUROPE

Ageing populations present many challenges, both to individuals and to society at large.[1] Ageing itself is, according to the Lund Declaration (Svedin 2009), a major societal challenge of our time, affecting issues of economics and social inclusion. As a result of declining fertility, increasing longevity, and the availability of better healthcare, population ageing is a phenomenon that a growing number of countries worldwide are experiencing. Globally, the number of people aged 60 and over will nearly triple in size, increasing from 894 million in 2010 to 2.43 billion in 2050 (UNPD 2010). In Europe, one quarter of the population will be over the age of 65 by 2020 and the current trend of early retirement is expected to put great strain on spending for pension, as well as health and long term care (McLean, 2011). In 2010, 36 million people worldwide were identified as having dementia, with this number expected to climb to 115 million by 2050 (Mihailidis et al. 2012). Dementia is just one instance of a cognitive or physical ailment associated with ageing, and one of the many challenges that age related technologies and policies must anticipate and address in their support of Europe's ageing population. Regardless of age, however, having a home is undoubtedly one of the most basic of human needs: the right

---

[1] | This work is a product of VALUE AGEING, Incorporating European fundamental values into ICT for ageing: a vital political, ethical, technological, and industrial challenge. A 48 month project funded by the European Commission-FP7 Marie Curie Industry-Academia Partnerships and Pathways Action. During the course of this project, expert interviews were conducted, referenced within this text.

to adequate housing is founded and recognised under international law. Described under article 25(1) of the Universal Declaration of Human Rights, the right to adequate housing is one that has also been identified within other major international human rights treaties. Referring to much more than the robustness of a building, "adequate housing" encompasses also the intangible, but no less essential elements of what makes a dwelling into a home. This includes creating a private space that is secure and safe, which encloses and facilitates the formation and maintenance of human relationships and personal bonds. Without proper support, older people experiencing diminishing cognitive or physical capacities may be forced to relocate to a central care centre, such as a care home, regardless of whether they desire to do so or not. Supporting those who wish to remain independent in their own homes as they grow older, who wish to age "in place", should be central to any strategy addressing the challenges presented by ageing populations, both for the benefit of older individuals, and society as a whole. As the term suggests "ageing in place" refers to the ability to grow old in one's home. Supporting this involves the design of ICT-based solutions and structural modifications in a home to meet the resident's needs, thus enabling them to remain in their homes for a greater length of time than would otherwise be possible. Fostering ageing in place is desirable, not only as it can improve the quality of life of those who wish to remain in their homes, but also because it can reduce the cost and the burden that age-related institutional care poses to society. Creating smart solutions and services in the home environment can dramatically improve quality of life for older individuals. The capabilities of the "smart home" concept are still expanding. This is partly due to the rapid advance of ICT and to the fact that a smart home is not a single device, but rather involves the use and integration of many different types of technology, as well as input from service providers. With respect to both housing design and smart homes, there is no "common user" and, to be fully effective in the solution proposed, each design must fully reflect the specific needs of the person. These should address internal factors – such as any mobility or sensory issues they may have –, as well as external factors, for instance the geographic location of the home or the penetration of broadband in the area, which is a crucial element in deciding what solutions are beneficial or even possible in each case. Finally, in deciding the best course of action, it is necessary to consider the physical capabilities of those who live in the house, and the human factors relevant to the design. Effectively supporting ageing in place means addressing many challenges, particularly in designing for the user's capabilities. This means ensuring that both the physical space of the home and the social fabric woven through it are accessible and open to the user, enhancing their independence and allowing them to live full, happy lives. In addition, older people in particular often live alone. Within the EU, just under half of older people over 65 years old (48.3 per cent) live as a couple, compared to 31.1 per cent living alone (Stula 2012). Although in some cases living alone is a deliberate choice, it can also be the result of the passage of time, such as the death of a spouse or the departure of grown children to homes of their own. Either way, for those for whom living alone is not a deliberate choice, smart and assistive technology can offer a way to carry on with their lifestyle and habits for longer. The home is more than bricks and mortar: it is the sum total of all meaningful social connections belonging to the residents. Considering the home in this manner reminds us that the needs we must address in supporting ageing in place extend beyond making the environment accessible; we must also support the psychological and social needs of the resident in order to maintain their well-being. In ancient Greek the word for the house was *oikos*. However, *oikos* can refer not just to the physical structure, but also to the social elements and the family unit, reflecting the social layer of home life. When understood as representing meaningful connections of people in the home environment, we can imagine the *oikos* of an older person living independently to include not just family members, but also the caregivers and regular visitors; those that interact with the older person on a regular basis and in person. In the case of older adults, the majority of these face-to-face interactions take place in the home, which becomes a social hub as well as a place to return when the day ends.

How can a smart home meet all of these needs without diminishing the social value of *oikos*? For the remainder of this chapter, we highlight the most pressing psychological and social implications that designers, researchers and policy makers must be aware of when considering the role of ICT in supporting ageing in place, and we will close with some proposed recommendations to assist in addressing the identified issues.

2. HOME REHABILITATION AND RE-STRUCTURING OF SPACE

The concept of *oikos* is represented in the WHO's understanding of housing, as represented in the 2006 report on housing and health regulations in Europe. The report, which examined the relationship between the quality of housing and the health status of residents, identified housing as not just a building, but the "conjunction of the dwelling, the home, the immediate environment and the community" (WHO 2006: 7). The WHO's "Global Age Friendly Cities" initiative aims to encourage cities and communities around the world to better meet the needs of their older populations The initiative relies on a WHO published document containing guidelines on how to be more open and accessible to older people across all aspects of the community, including outdoor space, transport and in promoting social participation. As part of making housing more accessible to older people, the guidelines make recommendations in nine different areas (WHO 2007): Affordability, Essential services, Design, Modifications, Maintenance, Ageing in place, Community integration, Housing options, and Living environment.

In order to support a person's desire to age in place, it is often necessary to rehabilitate the home, i.e. to adapt the home to account for the physical or cognitive decline of the resident. These guidelines highlight the need for the home to be altered to suit the resident's needs, such as having hallways and spaces wide enough for a wheelchair, or appropriately designed bathrooms and kitchens. Such physical alterations are intended to increase the accessibility of the home, with special consideration paid to human factors in design, i.e. the characteristics of people and their interactions with their environment to perform specific activities (Fisk et al. 2009). How the home is rehabilitated will depend greatly on the specific capabilities of the person and the age-related changes they experience: such changes can include a decrease or loss of fine motor function or mobility, visual or auditory impairments, and cognitive decline (memory loss, dementia etc.). For older people experiencing mobility difficulties, it may no longer be practical to have to climb stairs to use the bathroom, for example. Depending on the specific context, a chair lift may remove the barrier the stairs pose to the accessibility of the home, although if the resident is unable to operate the device, or if the environment does not allow for its installation, it may then be necessary to move the bedroom or install a bathroom on the ground floor. Ageing often increases the overall fragility of a person, meaning that falls are both more likely and more damaging should they occur (Rockwood et al. 1994; Ambrose et al. 2013). Reducing the risk of a fall may require relatively minor changes to the home: installing handrails or removing trip hazards such as loose rugs are two examples of how this issue may be addressed without major structural changes to the environment. Devices such as wearable or ambient sensors show promise in reliably triggering an alarm should a fall occur, and demonstrate how supporting ageing in place can be achieved through a combination of physical alterations to the home environment and careful implementation of ICT. Homelabs allow devices or alterations to be assessed in a homelike setting; recent initiatives such as the KUBIK project extend the concept. KUBIK is the product of collaboration between 28 international companies and is designed to allow companies to assess, not only minor alterations or ambient devices, but also light and wall placement, floor materials and many other environmental factors in a rapid and cost effective manner. Such approaches underscore the need to consider the impact of measures to support ageing in place, not just as individual alterations, but as a part of a larger integrated solution that serves the specific needs of the resident. According to the "Age Friendly Cities" guidelines, modifications to the home should not only be available but also affordable, and financial assistance should thus be provided to those that need it. In terms of ageing in place, the guidelines suggest that homes intended for older residents be located near services. In addition, these services should also be affordable and older people should be made aware of them, so that they can benefit from them. Assessing the cost effectiveness of smart homes for ageing in place is a complex task, particularly over longer periods of time (Magnusson/Hanson 2005); however, the cost of such a solution remains a concern for older people (Lê et al. 2012). The cost of smart home technologies (both the initial cost of a device and the ongoing cost of a subscription to a service) or the cost for the physical alteration of a house represent indeed a potentially serious barrier to their adoption. Essential services and suitable housing designs and modifications to support ageing in place need to be accessible to older people and, in this context, accessibility means that services are available, affordable and known.

3. ICT SOLUTIONS FOR AGEING IN PLACE

Smart homes are intelligent environments designed to increase the quality of life of the resident: for older people with diminished cognitive or physical capacity, smart homes have potential to increase their independence. The goal in developing smart homes for older people is twofold: to support their ability to age in place and to control health care costs, which will only increase as the average age of the European population increases (Demiris/ Hensel 2008). To meet this goal, smart homes must be equipped to increase the security of the resident and to reduce the risk of falls, stress, fear, isolation and a wide variety of other potential issues (Morris et al. 2013). As such, "smart home devices" refer to a range of technologies that widely vary in both form and function and that can be placed or embedded in the home with the goal of improving the quality of life of the user. Within Europe, the Ambient Assisted Living Joint Programme (AAL JP), a European Commission financed funding activity initiated in 2008, seeks to encourage social, technological and business innovation to achieve:

- New models of service delivery and care that contributes to greater self-reliance for older adults and greater support for informal carers;
- Adapted living spaces that can improve the quality of their everyday lives;
- New ways for older people to remain active, including contributing as volunteers or providing mutual support;
- New ways of mobilising active and trusted networks, both formal and informal, professional and in kind, to provide all types of support.

The projects funded under the AAL JP illustrate the spread of technologies and functions that can be considered under the label of "smart home technologies", as well as the range of needs that technology may address. While some projects looked towards embedded or ambient sensors to assist in fall prevention (CARE, HOPE), others approached the same need addressed by way of wearable sensors (A2E2). Other AAL JP funded projects have directed their attention toward robotics (EXCITE, ALIAS), social networks (JOIN-IN, CVN, AWARE, OSTEOLINK), online services for connecting with care givers, health professionals or community groups (V2ME, SI-SCREEN, NOSTALGIA BITS), and serious games which provide older people with mental and physical exercises (SILVER GAME, GAMEUP, MOB MOTIVATOR). Smart home devices may be employed to address a specific need – for instance, to trigger an alarm in the event of a fall – or they may be part of a network with the broader goal of enhancing the user's quality of life. This can be achieved by increasing the security of the home (for example, sensors which trigger alarms if they detect smoke or water leakage, or which automatically turn off an appliance after a set amount of time), or by enhancing the independence of the user by assisting them in taking the correct medication at the correct time. Smart home technologies also have a commercial presence in Europe. Networked or "smart" devices already proliferate in the home, and each year sees companies push further into health related technologies for the home. Products such as the Nest Thermostat™, which can be controlled remotely via a smartphone, and the SmarterKey™ lock, a keyless lock which, again, can be controlled via a dedicated smartphone app, represent recent examples of older aspects of the house being re-conceived as modern appliances, with a view toward making home life more energy-efficient or more convenient. In addition to enabling users to better manage home utilities, smart devices are also envisioned as a means to give users more control over their own health. The Aria™ WiFi Smart Scale, for instance, tells the user more than just their weight, it also claims to track the user's body fat percentage and BMI, tracking these data over a period of time and connecting to a computer or smartphone to allow them to track progress and set fitness goals. Such smart devices give a window into the world of the "internet of things", where household objects interact with each other and provide the user with data, or even gamified "objectives" to incentivise certain behaviours (eat more of a certain kind of food, run a certain distance per week, etc.). However, while discrete smart devices are increasingly a commercial normality, the fully integrated smart home concept is yet to appear on a large scale. In addition to initiatives such as the AAL JP, groups such as Smart Homes, a Dutch consultancy which specialises in promoting and assisting in the research of automated or smart home concepts, seek to champion their potential as effective eHealth and energy management tools in the future and help to create an environment where smart homes can become a viable and sustainable aspect of home life in the near future. Smart homes are intended to enhance the independence of the user by providing unobtrusive support systems that can meet their needs (Morris et al. 2013): in the case of older people, this can mean addressing the symptoms of cognitive or physical decline. However, there are other aspects of life that designers and policy makers must be aware of, and the effects of technology in the home can reach far beyond the original or intended goal.

4. ADDRESSING THE SOCIAL AND PSYCHOLOGICAL DIMENSIONS

The goal of the technologies listed above is to improve the quality of life of the user. This goal is often translated into addressing the diminishing capabilities – both physical and cognitive – of the user, enhancing their independence. User specific limitations and needs are taken into account during the design of smart technology through well-known design approaches like the User-Centered Design and the task based design methodology (Rosson/ Carroll 2002; Fisk et al. 2009). However, there are other dimensions that should be considered when designing technology for ageing in place. Assessments of the quality of life of older people are largely health-focused, with indicators considering functional capacity, health status, psychological wellbeing, social support, morale, dependence, coping and adjustment, all of which are used as "proxies" for Quality of Life (QoL) assessment (Walker 2005). While these indicators cover a broad spectrum of issues relevant to the quality of life of older people living at home, they also limit comprehensive evaluations by not considering the subjective perspective of the older person whose quality of life is under evaluation. The shift toward recognising the quality of life of older people as a multi-dimensional concept means incorporating both objective and subjective aspects alongside expert assessments informed by the opinions and feedback of older people, and is reflected in the development of new assessment tools, such as the Subjective Evaluation of Individual Quality of Life (SEIQoL). With SEIQoL for example, the relative importance of individual factors affecting quality of life is not fixed, but established on a case-by-case basis through the use of interviews and visual aids, such as moveable pie charts (Barnes 2002). Supporting ageing in place is about meeting the needs of older people and enhancing their Quality of Life, and that this is increasingly seen as multifaceted, with more consideration given to subjective assessment. As a result, it is not sufficient to consider only the physical or purely objective health outcomes in the implementation of smart home technologies. The following issues, grouped under the umbrella of psychological and social dimensions of smart home assisted ageing in place, must also be considered and accounted for by all stakeholders involved in the creation of home solutions:

- Social isolation and loneliness
- Sense of identity
- Privacy
- Control
- Sense of identity
- Frustration and abandonment
- Acceptability of measures to support ageing in place.

Social Isolation and Loneliness

One of the greatest challenges to supporting ageing in place is the need to address social isolation and loneliness. A 2010 Eurostat study, "Social Participation and Social Isolation", found that the risk of social isolation increased as people aged (Lelkes 2010). This trend was found to be present to varying degrees across the EU. Half of the countries surveyed reported that at least 10 per cent of respondents over the age of 65 had no contact with friends, either in person or remotely. In Hungary and Lithuania, this figure rose to 25 per cent. With the exception of the Netherlands and Denmark, the study also reported that although the elderly are strongly affected by diminishing interaction with friends or relatives, "in many countries they can still rely on the help of others, to about the same extent as their younger compatriots" (ibid: 33). The findings of the Eurostat survey are valuable in that they provide an image of the pervasiveness of social isolation among older populations across Europe.

Addressing social isolation, particularly on a large scale, poses many difficulties as it can be the result of one, some or all of series of different factors connected to all aspects of the older person's life. Physical frailty, lack of finance, the location of the home, and changes in the older person's social circle (both family and peers) can all contribute to a greater or lesser degree to people being isolated from their community and society at large. Loneliness can be a product of this isolation and is reflective of dissatisfaction with the frequency or closeness of social relationships they have, when compared against what they would like to have (Steptoe et al. 2013). Social isolation has social, psychological and even physical implications for those who experience it, particularly those in need of care and assistance. A person is deemed to be isolated or not based on factors such as contact with other people, or the physical location of the home, while the feeling of loneliness is one of the psychological implications of isolation and has been associated with higher mortality in older men and women. Although it can be a symptom of cognitive or physical decline, loneliness is not in itself an illness: however, it does play a large role in a person's well-being and quality of life. Unlike social isolation, loneliness is largely subjective in that a person may have little contact with others throughout and be perfectly content. In contrast, social isolation can be assessed by evaluating frequency and quality of contact with others, ease of access to services and transport, ability to participate in social events, etc. Recent studies have found that, while loneliness did not have a measurable, direct impact on the mortality of older people (Steptoe et al. 2013), it can increase the likelihood or rate of cognitive decline, with older people classed as "lonely" more than twice as likely to develop dementia (Wilson et al. 2007). While studies such as those by Eurostat (discussed above) illustrate the need to address social isolation and loneliness as a common need shared by older people across Europe, large scale policy or technological solutions alone cannot effectively address the challenge this need presents. Although older people across Europe may share a common risk of becoming isolated or lonely, it is a destination reached through many roads. That is not to say, however, that there is nothing to be achieved through careful and innovative implementation of ICT and smart home technologies. Services such as email have been available for some time, but the increase in the availability of video conferencing capabilities in home computers, phones and tablets allows for real time communication, with visual as well as auditory feedback to the users. ICT serves a social need by connecting family and friends, but to do so it must be available and accessible to all. It also should be viewed as a tool to enhance, not replace, human contact. In addition, while they may not provide the same level or quality of social interaction, services such as online shopping with delivery do allow older people to retain agency over an important part of daily life, restocking the home. This is, however, not without its own challenges, as it depends on the design of online shopping service being accessible, both to those with physical impairment (e.g. can the text be resized or recoloured for those with visual impairment) and for those with a lower level of digital literacy or less experience accessing services online (e.g. does each listed item have an image so the user can be sure about what s/he is purchasing?; is the service complicated to use without a high degree of digital literacy?). However, to those whose isolation is in part the result of deteriorating financial conditions, cost may prove to be a barrier to accessing technological solutions and, as such, they will not assist those experiencing isolation or loneliness without effective policy measures or innovative social strategies (e.g. a bus service that collects people from their homes and brings them to the town centre or market on specific days, or community groups that actively involve those that may otherwise grow isolated from those around them).

"Successful" implementation of ageing in place strategies requires that social isolation be addressed, and smart home technologies have the potential to be a powerful tool in doing so. Network technologies make it possible to communicate with people in ever more new and varied ways, allowing easier access to information, and to other people. However, addressing needs such as loneliness and social isolation requires a sensitive and deliberate approach to implementing technology-based solutions in the homes of older people. Although these needs are commonly shared across Europe, the root causes behind it can vary wildly across individual instances and installing smart home technologies may fail to address them or – worse – exacerbate the situation and have a negative impact on the person's quality of life.

Sense of Identity

As people age the amount of time spent in their home increases (Walker 2005): the home is reflective of the individual's experiences and personality, which over time is imbued with its own identity. This identity is drawn from tradition, social cohesion, history, even through deliberate manipulation or alteration of the space to suit the needs and preferences of the occupant. As such, care must be taken when modifying the home to support ageing in place. The importance of a sense of identity in the environment is well studied with respect to care homes and is equally important in maintaining a sense of wellbeing with ageing in place (Robichaud et al. 2006; Oswald et al. 2007). This sense of identity is important, not only since preserving it reduces the risk of frustration and increasing the acceptability of adapting the home, but also because of the numerous positive implications on older people, particularly those with cognitive or physical impairments. Difficulties older people may have with activities of daily living are reduced when performed at home or in spaces they are very familiar with. Familiarity with the environment and its features enables those who are frail or suffering from sensory and physical decline to navigate spaces that would normally require greater effort. People with dementia are also documented to be able to perform tasks otherwise outside of their capabilities when performed within their own home (Mihailidis et al. 2012). Supporting ageing in place can involve modifying the home through physical alterations and the addition of smart home technologies, but attention should be given to retaining the familiarity of the home space for the occupant, preserving the identity they have instilled upon it. Physical modifications – such as replacing steps with ramps or installing support rails in the bathroom – do not alter the layout or function of rooms in the home, yet make it easier to complete daily tasks independently. Considered use of both physical alterations combined with smart home technologies can greatly increase the independence of the resident, allowing them to remain in their homes for longer without drastically altering the layout or physical environment. One of the most valuable aspects of the home is that it creates a distinction between the public and private space: this distinction not only enhances the sense of independence of the occupant, but is also fundamental in promoting a sense of identification in users (De Matteis 2010). This is particularly important when modifying the home and involves the installation of monitoring devices or ambient technologies. Eroding the distinction between public and private spheres may also inadvertently dilute the sense of identity, familiarity and comfort the person has within the home. Even passive technologies, which may not take up much space, can be deemed unacceptable by residents if they view the ability of the devices to monitor or record within the home as intrusive (Guihen 2013). Introducing assistive measures in the home can make life for the resident easier and performing daily tasks less daunting. However, they may also alter the environment in ways that leave the resident feeling uneasy, or less "at home". This concern is particularly valid in the context of smart home technologies, the functions of which may not be immediately clear to the user, causing distrust or suspicion. In particular, care must be taken to ensure that monitoring devices do not erode the distinction between public and private space but instead find a balance between the two, so that the resident feels neither shut away from society nor completely exposed and without privacy. This balance is vital, as it plays directly into the resident's feelings of safety and security in the home, while it affects their acceptance and receptiveness to support measures that enhance independence and allow them to remain in their homes.

Privacy

In Europe, the right to privacy is recognised as fundamental. As technology grows increasingly networked and our data continue to move online, it is vital that this right be safeguarded. Privacy also involves the shielding of people from unnecessary or excessive categorisation, which can have a negative impact on the quality of life of older people (De Hert/Mantovani 2010). All smart home technologies require data to be effective, be it an ambient sensor that can detect when someone is in the room, a wearable device that logs movement, or software which tracks the daily progress made in cognitive or physical exercises. As such, it is vital that privacy issues be assessed from the earliest concept designs and all the process through to implementation. This is especially true in the context of medical health data, which smart home technologies can be used to collect, store or even transmit to other locations, such as a hospital or doctor's surgery. Improper handling of medical or user data can expose the user to unsolicited contact and potential harm (De Hert/Mantovani 2010). Even though monitoring is a function of smart home technologies and a potential benefit for assisting ageing in place, one must be careful to ensure that the privacy and autonomy of the resident is respected. In assessing the functional capacity of a device to meet the needs of the resident, for instance, questions must be asked with respect to its effect on the privacy of the resident. For example, although cameras may provide a great deal of information in the event of a fall, a study on the attitudes of older people toward smart home technology conducted by the University of Missouri-Columbia found that such a measure was universally felt to be "obtrusive", with concerns raised as to the potential for privacy violations. The same respondents were more favourable to the technology when data could be obscured or "anonymized" (Demiris/Hensel 2008). Smart home technologies are intended to meet the needs of the user, but for them to be deemed acceptable, privacy concerns must be addressed. This study is not alone in earmarking cameras in particular as a barrier to trust and adoption of smart home technologies. More recent reviews continue to highlight such devices as a concern, and privacy as the primary barrier to the adoption of smart home technologies among older adults (Morris et al. 2013).

Control

One of the benefits of AAL technology is that it can act independently of active user input, triggering an alarm in the event of a fall, for example. AAL technology, such as sensors for monitoring motion, heat, electricity usage etc., is designed to operate in the background throughout the day. However, while this may enhance the effectiveness of the technology, it can also have a significant psychological impact on the resident. Smart home technology can be divided into two groups: active and passive (Guihen 2013). Active technology includes devices such as telephones, user triggered alarms and controls for house utilities such as heating. Passive technology does not require the user to actively engage with it and includes, for example, house sensors. The issue of control and the level of control provided to the resident is one of the most important factors influencing user acceptability of smart home technology. While passive technologies have the benefit of working irrespective of the abilities or interest of the user, it is this exact feature that can lead older people to prefer engaging with active technologies, as they offer a more tangible sense of control (Guihen 2013). Unlike an ambient sensor placed discreetly in the corner, active technologies provide a sense of agency to the users, as they are the ones directing how and when it is used and – crucially – not used. The issue of control is strongly connected to that of privacy and personal space. The user wants to be in control of their environment, making the technology work for them and toward their preferences rather than living in an environment where devices monitor or record their movements and actions without any active feedback for the user. As a result, implementing smart or ambient technologies must be done in a manner that does not remove the users' sense of autonomy and control over their own home. Identifying the appropriate level of control is particularly difficult in cases where the resident suffers from dementia or Alzheimer's disease. During the interviews conducted as part of the VALUE AGEING project, Rafael Capurro, suggested that "In case of degenerative diseases such as Alzheimer's the balance goes towards taking the place of the other by others and/or by technology" (Guihen 2013: 58). In such instances, it may reduce the capability of the technology if too much is asked of the user, and a higher level of automation may be necessary. However, such decisions must be made on a case-by-case basis, with the enhanced autonomy and safety of the older person in the forefront of the decision-making process. The divide between automated and directly controlled is not impermeable and there may be instances where the hybrid of both approaches is most appropriate. For instance, while ambient sensors can monitor the environment and react to changes within it based on preset parameters; they need not necessarily actuate another element of the smart home platform. Instead of raising an alarm or flagging the attention of a caregiver, the device may prompt the user to take action directly. Another interviewee, Alex Mihailidis, is optimistic about such approaches: "I think the technologies that are able to monitor older adults during common activities, provide them with prompts and reminders of things they need to do, systems that can monitor the health and safety of older adults, such as fall detection systems or ones that collect some very basic, high-level physiological health data to the person and make that available to the older adults themselves and their caregivers; those are the main classes that I would see as the most promising in the short term right now." (Guihen 2013: 40) The COACH[2] project is indicative of this approach.

---

2 | Cognitive Orthosis for Assisting with aCtivities in the Home, a prototype of an intelligent environment designed to assist people with dementia manage activ-

It developed a Cognitive Assistive Technology (CAT) designed to remind and assist older people in washing their hands before they leave the bathroom. CATs leverage smart home technologies to increase the independence of older adults by alerting the user that an action is required. However, the autonomy of the user and their sense of control over the home environment are preserved by prompting the user to act rather than completing tasks independently. This is particularly beneficial in instances where the resident is suffering from mild cognitive decline and may forget to perform basic tasks around the house (Mihailidis 2008). Providing the resident with greater control over their environment while still providing the assistance necessary to enhance their independence can be a difficult balance to achieve. However, such a balance is necessary to ensure that the resident feels at ease in their own home, that is, supported by the technology rather than controlled by it.

Frustration and Abandonment

Even as the capabilities of smart home technologies continue to advance, older users may not be enthusiastic to adopt them into their lives. There is also the possibility that the user may abandon the technology, completely removing any potential benefit it may have had. One of the greatest reasons for user abandonment is frustration (Guihen 2013): frustration occurs when the device or service fails to meet user expectations, or is found to be difficult to use. Such frustrations are of particular concern with emerging technologies, where the user may have no prior experience with a device and no initial comfort can be assumed. The inability of the user to adapt the device to their specific needs – either as a result of inflexible design or poor training – increases the risk of the user growing frustrated with the technology either actively choosing to abandon it or using it with less frequency as time goes on. Frustration with a technology can also be the result of discomfort felt by the user, either as it is uncomfortable to use (such as a bulky wearable sensor, or a device with a confusing user interface) or because the user experiences an alienating effect, such as can happen when they do not feel in control of devices in their own home. Frustration is also not limited to ICT based solutions: for instance, a ramp installed in place of the front step to reduce the risk of fall may be uncomfortably steep for the user or a metal handrail too

---

ities of daily living in their homes with greater independence, managed by the Intelligent Assistive Technology and Systems Lab at the University of Toronto.

cold for comfortable use in the winter. Such discomforts may also contribute to frustration felt by the resident and, while one cannot switch off a front step, this frustration can contribute to diminishing the comfort felt by the resident in their own home.

Acceptability of Measures to Support Ageing in Place

There is currently no formal ethical framework for assistive technologies (Tiwari et al. 2010): however, work has been carried out to develop guidelines and conditions for the successful and ethical development of assistive technologies. The SENIOR project, a 24 month European Commission funded project tasked with assessing social, ethical and privacy issues associated with ICT and ageing, identified several conditions that should be addressed when developing technologies to support older people: these conditions influence not just individual acceptance of a technology, but also societal acceptance. SENIOR highlighted the needs for user freedom of choice to be respected, as well as their autonomy and privacy. It also recommended that any solution implemented safeguard the dignity and self-esteem of the user, while emphasising safety (Wright 2009). Carefully assessing needs and matching these to the provided technology can result in higher acceptance rates. In an attempt to study elderly needs toward the adoption of assistive technologies, Forlizzi, DiSalvo and Gemperle (2004) conducted an ethnographic study of ageing adults who lived independently in their homes (seventeen elders aged 60 through 90 were interviewed and observed in their homes). Self-awareness and self-perception of the user's own needs have been showed to affect the way they feel towards assistive technology, increasing the likelihood that they will use technology once they identify or understand the potential benefits of the proposed device. Older people may recognise the utility of a technology, but may not feel that it could assist them specifically, nor do they desire to use a technology that they feel exaggerates or draws attention to their needs (Guihen 2013). The relationship of older adults with technology cannot be compared with the relationship that younger adults experience: they may not be as familiar with ICT terminology, in particular with "interface metaphors". While younger adults are familiar with terms such as mice, folders, and desktops, as they relate to technology, these concepts may be unfamiliar to older users, who are then required to acquire a new lexicon of ICT related vocabulary (Guihen 2013). Such examples are illustrative of how the barrier to entry with technology may inadvertently be raised, even from the early design phase. Care must be taken to be aware of such issues and to address them from an early stage to minimise the risk of reducing the acceptability of smart home technologies in the eyes of the end user.

5. LOOKING TO THE FUTURE OF AGEING IN PLACE IN EUROPE

Europe's ageing population is a social reality: however, that does not mean we cannot take steps to address the issues this presents to us. Supporting ageing in place with smart technology and home rehabilitation has the potential not only to reduce the societal cost of long term institutional care, but also to empower older members of our communities to remain independent in their homes for longer and with a higher quality of life. However, the potential for technology to address the needs of older people in this context also brings with it the temptation to see these devices and services as a panacea to the challenges of fulfilling the desire of older people to remain in their homes. This chapter has highlighted needs which fall outside the traditional understanding of physical and cognitive challenges associated with ageing: social isolation and loneliness, privacy, control, sense of identity, frustration and abandonment, and the acceptability of measures to support ageing in place. To conclude, we will suggest some recommendations intended to ensure that these social and psychological needs are addressed, increasing the acceptability of smart home technologies for ageing as well as the overall quality of life of those that choose to use them. Within Europe, our core values are enshrined in the European Charter of Fundamental rights. We recognise the vital importance of ensuring that the dignity, privacy and security of every person are preserved. We need to identify the values we wish to see embedded in housing design and smart homes: these values are not simply abstract concepts and should inform guidelines to which we adhere when addressing societal challenges. As such, we should develop tools to help us ensure these values are embedded effectively and to assist us in assessing the implications so that further refinements can be made as time moves on. The WHO Guidelines, the Riga Declaration and Digital Agenda for Europe are examples of how values can be operationalised and translated into policy goals. Identifying stakeholders is a vital part of addressing the needs of all involved in supporting ageing in place, including designers, formal and informal caregivers, policy makers and the users themselves. However, it is not enough to identify them; efforts should be made to facilitate communication between them and to increase the impact of dissemination efforts. The AAL JP, for instance, brings stakeholders together and incentivises efforts focused on specific identified societal needs, such as ageing in place or assisted navigation in public spaces. It is also a platform for dissemination, both of best practice efforts and new technological innovations. The barriers to proposed solutions must be identified and mitigating strategies proposed. In particular, we must ensure that the benefits of smart home technologies are shared by all and that issues such as cost or the location of the home do not become insurmountable barriers, keeping new technology and design innovations from benefiting the most vulnerable in society. Finally, we need to be aware that new practices, norms and technologies will emerge which will challenge existing knowledge. Much as we should not only be receptive to but also seek out cross-stakeholder collaborations and end user feedback, we must also be willing to revisit and reevaluate existing best practice norms in the face of shifting societal paradigms. The needs and responsibilities identified in this paper should be placed in the context of the European Charter of Fundamental Rights and the Values of the European Community applied to address them. If appropriate measures are taken to ensure that the needs of older people and their associated *oikos* are at the centre of the development of new policies and smart home related technologies, their potential for societal and individual good may yet be met.

REFERENCES

Ambrose, A.F./Paul, G./Hausdorff, J.M. (2013): "Risk factors for falls among older adults: A review of the literature." In: Maturitas 75, pp. 51-61.

Barnes, S. (2002): "The design of caring environments and the quality of life of older people." In: Ageing and society, 22, pp. 775-789.

De Hert, P./Mantovani, E. (2010): "On Private Life and Data Protection." In: Mordini, E./De Hert, P. (eds.), Ageing and Invisibility, Amsterdam, The Netherlands: IOS Press.

De Matteis, F. (2010): "Housing Europe." In: Clemente, C./ De Matteis, F.,(eds.): Housing for Europe: Strategies for Quality in Urban Space, Excellence in Design, Performance in Building. Rome, Italy: The Urbact II Operational Programme 2007-2013, Working Group HOPUS- Housing Praxis for Urban Sustainability.

Demiris, G/Hensel, B.K. (2008): "Technologies for an Aging Society: A Systematic Review of "Smart Home" Applications." In: Geissbuhler, A./ Kulikowski, C. (eds.), IMIA Yearbook Of Medical Informatics 47/1, pp. 33-40.

Fisk, A.D./Rogers W.A./Charness N./Czaja S.J./Sharit J. (2009): Designing for older adults. Principles and creative human factor approaches. (2nd ed.), Florida, USA: CRC Press LLC.

Forlizzi, J./DiSalvo, C./Gemperle, F. (2004): Assistive robotics and a ecology of elders living independently in their homes. Human-Computer Interaction Institute.

Guihen, B. (2013): Human Issues related to housing design for elderly and smart homes, report for FP7 Marie Curie IAPP Project "VALUE AGEING", Grant agreement: 251686, http://va.annabianco.it/wp-content/uploads/2014/09/07AGE03_D3.5_FINAL.pdf . Last access: 03.11.2014

Lê, Q./Nguyen, H.B./Barnett, T. (2012): "Smart Homes for Older People: Positive Aging in a Digital World." In: Future Internet, 4/2, pp. 607-617

Lelkes, O. (2010): "Social participation and social isolation". In: Atkinson, A.B./Marlier, E. (eds.), Income and living conditions in Europe. Eurostat Statistical books: Luxembourg: Publications Office of the European Union, pp. 217-240.

Magnusson, L./Hanson, E.J. (2013): "Ethical issues arising from a research, technology and development project to support frail older and their family carers." In: Health and Social Care in the Community 11/5, pp. 431-439.

McLean, A. (2011) "Ethical frontiers of ICT and older users: cultural, pragmatic and ethical issues.", In Ethics and Information Technology, 13(4), pp. 313-326

Mihailidis, A./Boger, J.N./Craig, T./Hoey, J. (2008): "The COACH prompting system to assist older adults with dementia through handwashing: An efficacy study." In: BMC Geriatrics 8/28, http://www.biomedcentral.com/1471-2318/8/28. Last access: 03.11.2014

Mihailidis, A./Boger, J./Czarnuch, S./Nagdee, T./Hoey, J. (2012): "Ambient Assisted Living Technology to Support Older Adults with Dementia with Activities of Daily Living: Key Concepts and the State of the Art." In: Augusto, C./Huch, M./Kameas, A./Maitland, J./McCullagh, P./Roberts, J./Sixsmith, A./Wichert, R. (eds.), Handbook of Ambient Assisted Living: Technology for Healthcare, Rehabilitation and Well-being, Amsterdam, The Netherlands: IOS Press.

Morris, M.E./Adair, B./Miller, K./Ozanne, E./Hansen, R./Pearce, A.J./Santamaria, N./ Viegas, L./Long, M./Said, C.M. (2013): "Smart-Home Technologies to Assist Older People to Live Well at Home." In: Aging Sci 1:101.

Oswald, F./Wahl, H./Schilling, O./Nygren, C./Fänge, A./Sixsmith, A./Sixsmith, J./Széman, Z./Tomsone, S./Iwarsson, S. (2007): "Relationship Between Housing and Healthy Aging in Very Old Age." In: The Gerontologist, 47/1, pp. 96-107.

Robichaud, L./Durand, P.J./Bédard, R./Ouellet, J. (2006): "Quality of Life Indicators in Long Term Care: Opinions of Elderly Residents and Their Families." In: Canadian Journal of Occupational Therapy October 73/4, pp. 245-251.

Rockwood, K./Stolee, P./Duncan, R./Beattie, B.L. (1994): "Frailty in elderly people: an evolving concept." In: CMAJ 150/4 pp. 489-496.

Rosson, M.B./Carroll, J.M. (2002): "Scenario-Based Design." In: Jacko, J./Sears, A. (eds.), The Human-Computer Interaction Handbook: Fundamentals, Evolving Technologies and Emerging Applications. New Jersey, USA: Lawrence Erlbaum Associates, pp. 1032-1050.

Steptoe, A./Shankar, A./Demakakos, P./Wardle, J. (2013): Social isolation, loneliness, and all-cause mortality in older men and women, Proceedings of the National Academy of Sciences of the United States of America, 110/15, pp. 5719-5801.

Stula, S./Müller, S. (2012): Living in Old Age In Europe- Current Developments and Challenges, Observatory of Sociopolitical Developments in Europe, (Working Paper no.7), , http://www.sociopolitical-observatory.eu/uploads/tx_aebgppublications/AP_7_EN.pdf. Last access: 02.11.2014

Svedin, U. (2009): Final report from the New Worlds – New Solutions. Research and Innovation as a Basis for Developing Europe in a Global Context. Lund, Sweden: The Swedish Research Council for Environment, Agricultural Sciences and Spatial Planning (In consultation with the Lund Declaration Group), https://www.vr.se/download/18.29b9c5ae1268d-01cd5c8000631/New_Worlds_New_Solutions_Report.pdf. Last access: 02.11.2014

Tiwari, P./Warren, J./Day, K./McDonald, B. (2010): "Some non-technology implications for wider application of robots assisting older people." In: Health Care and Informatics Review Online, 14/1, pp. 2-11.

United Nations Population Division (UNPD) (2010): Department of Economic and Social Affairs, World Population Prospects: The 2010 Revision, http://esa.un.org/wpp/documentation/pdf/WPP2010_Volume-I_ Comprehensive-Tables.pdf. Last access: 03.11.2014

Walker, A. (2005): Growing Older in Europe. Berkshire, England: Open University Press.

Wilson, R. S.,/Krueger, K.R./Arnold, S.E./Schneider, J.A./Kelly, J.F./Barnes, L.L.//Tang, Y./Bennett, D.A. (2007): "Loneliness and Risk of Alzheimer Disease." In: Arch Gen Psychiatry 64/2, pp. 234-240.

World Health Organisation (2006): Housing and health regulations in Europe, Final Report, http://www.euro.who.int/__data/assets/pdf_file/0004/121837/e89278.pdf. Last access: 02.11.2014

World Health Organisation (2007): Global Age Friendly Cities: A guide. France: World Health Organisation

Wright, D. (2009): D4.1 Report on Best Practices and Roadmap towards the Roadmap, SENIOR Project, Grant agreement: 216820, http://globalseci.com/wp-content/uploads/2010/01/D4.1-Report-on-Best-Practices-and-Roadmap-towards-the-Roadmap.pdf. Last access: 01.11.2014

Seeing Again. Dementia, Personhood and Technology

IKE KAMPHOF

Too little is known about how telecare applications actually work, in particular about what they mean for the lives and identities of their direct users. This paper investigates one application, lifestyle monitoring technology, as it is used in homecare for frail elderly people. By bringing together a post-phenomenological approach with results from ethnographic research into Dutch homecare, I offer a more nuanced picture than the hopes and fears currently articulated around telecare. I argue that technologically mediated processes of observing vulnerable homecare clients require an intricate combination of human and technological "seeing" and "not seeing" to secure respectful care.

1. PERSONHOOD AND SMART LIVING

Homecare organizations employ lifestyle monitoring technologies – often still in a semi-experimental stage – to keep a watchful eye on frail elderly people living alone, in particular people with dementia. These systems provide data about patterns of behavior in the home and function supplementary to care that requires the physical presence of caregivers, such as help with washing or dressing and the administration of medicine. By detecting emerging health and safety hazards at an early stage, monitoring allegedly supports the independence and wellbeing of homecare clients (cp. Coronato/De Pietro 2010: 27-29; Ni Scanaill et al. 2006: 549; Price 2009: 12). However, monitoring systems are criticized for their vast surveillance power "where the person disappears and their body is coded as a data node" (Kenner 2008: 265; cp. Brittain et al. 2010: 98). Both these claims are largely speculative, based on an analysis of current care practice and technological possibilities.

The structure of technologies however is "multistable" (Ihde 1990: 144) and devices acquire their meaning in concrete contexts of use. Moreover telecare technologies do not leave these contexts untouched: they mediate care relationships, often in unforeseen ways, and transform the meaning of care. How this happens and specifically what it means for users on a daily basis is as yet in need of further research (cp. Pols 2012: 13; Bowes et al. 2012: 20).

User perspectives can be collected by, among others, organizing questionnaires or focus groups in which people are asked about what they think and want. But users – whether managers, caregivers or clients – often have no insight yet about what technologically mediated care means, even as they are involved in it. Qualitative ethnographic research of situated practices can help to develop an informed perspective that supports further discussions on telecare (cp. Pols 2012: 15). Ethnographic work is not neutral, nor does it result in mere factual description; it sounds out patterns that can have relevance beyond the scope of the case at hand. It also requires a structure to guide its own observations: that structure is here provided by a focus on the intimate relationships of bodies and their technological "extension[s]" (Ihde 1990: 40) as addressed by post-phenomenology. This perspective is particularly suited to people with dementia who cannot always communicate needs or express who they are by linguistic means, while bodily expressions of identity often linger (cp. Hughes et al. 2006: 173-176; Kontos 2005: 557). Additionally, I depart from three main qualms, raised in recent literature, about what lifestyle monitoring could mean for the identity of vulnerable people. I will now introduce these qualms in general and more specifically.

As Kitwood (1997) argued, our identity as persons is bestowed on us by others; this is particularly true for people with dementia who become increasingly dependent on other people. Kitwood criticized the dominant framework of the biomedical model of dementia as a brain disease that focuses on the mental decline that comes with dementia. Where our mind is often considered the core of our existence as independent, self-directing individuals, dementia tends to be portrayed as involving a loss of self. This depiction effectively makes people with dementia invisible as persons and easily leads to a "malignant social psychology" (Kitwood 1997: 4) that further undermines their personhood by stigmatization, infantilization and objectification. In contrast, care as "positive person work" (69), based on recognizing others as unique beings with their own abilities, actively supports personhood.

Kitwood's social interactionist view on identity can be fruitfully extended to other frail people, but as Kontos (2005: 555) rightly states, it disregards embodied aspects of selfhood. Interactionist, relational views on identity and views that focus on our being as embodied, however, do not exclude each other. The embodied self is a situated self, whose identity is constituted in interaction with both its social and material environment. This combined perspective suits the analysis of monitoring technology well: using motion sensors and occasionally camera's, monitoring systems register body movement in the home and interactions with objects as a person goes about their daily tasks. They turn personal living spaces into *smart* spaces that alert caregivers of emerging problems and partly serve to postpone a disruptive move into residential care. By this they affirm the role of the home as an extension of the self implied in current ideals of aging-in-place.

Monitoring also infringes on the privacy of the home by gathering extensive sets of data that are relayed and evaluated elsewhere. The *smarts* these technologies offer do not just compensate losses entailed in processes of de-menting, but also aim to relieve physical and mental burdens of informal and formal caregivers by *smart care*. On the level of healthcare policy, *smart living* is presented as a means to safeguard the healthcare system from the managerial and financial threats that the aging population is perceived to entail. Monitoring systems take cared for and caregivers up in "compounds", as Haraway (2008: 250) designates the extended networks that connect organic bodies, their various activities and technologies. Within given care compounds, caregivers and people cared for, technologies, homes and care policies and practices actively shape each other.

With regard to dementia care as "person work" (Kitwood 1997: 69), the use of lifestyle monitoring raises critical issues (cp. Baldwin 2005; Kenner 2008; Mahoney et al. 2007; Bowes et al. 2012). Placing monitoring within a sociopolitical context, Kenner (2008) argues that these systems function primarily as a tool for care management, translating behavior "into data about the body that may then be analyzed, categorized and regulated at a distance" (253). Three main interrelated concerns can be identified:

1. *The rule of norms.* The use of monitoring technology affirms the ageism in our society by approaching aging and dementia mainly as diseases: the behavior of vulnerable people is judged against models of healthy 'normality'. Focusing on biomedical pathologies easily leads to disregarding individual differences between people with dementia (cp. Mahoney et al. 2007: 220).
2. *Risk and control.* Within an ageist framework, frail elderly people are being seen as "at risk", which legitimates intrusion into their personal

space: monitoring redefines care as risk management and behavior control (Kenner 2008: 262). Baldwin (2005) raises the question how a person with dementia is reconstituted in relation to technologies that function automatically and override the person.
3. *Erosion of care relationships.* Kenner (2008: 257) rightly points to the discourses of fear around aged care that overshadow the needs of actual people. The focus is on finding a "technological fix" for the burdens of care, both on the level of management and the daily work of care. Baldwin (2005) suggests that the focus of technology on accuracy and efficiency may erode the quality of care relationships.

These concerns articulate important warnings that technologically mediated observation may confirm the malignant ways of seeing and treating frail people that Kitwood (1997) identified. However, in her focus on sociopolitical inequalities and individual rights, Kenner (2008) excludes the possibility that care giving, whether it is done by humans or by technology, can also stand in and work for vulnerable persons. Bowes et al. (2012) emphasize instead that "processes of care are processes of co-production" (14) between professionals, family and clients. Baldwin (2005) and Mahoney et al. (2007: 220) rightly raise the question of how technology mediates interdependencies within care relationships.

Homecare interdependencies play out at different levels. My concern here will be with the daily work of caregivers, instead of with the national and international discourses on the management of healthcare for elderly people. Though their work is influenced by the political-economical context, caregivers are guided less by management interests and more by a practical "logic of care" as a day-to-day "tinkering" (Mol 2008: 12) to improve or maintain their clients' condition.

A pervasive intuition in recent debates on personhood closely associates perceptive attention to the needs of frail elderly people with ethical respect for their unique personhood (cp. Hughes et al. 2006: 1-36). Respect comes from the Latin verb *respecere*, which means "to see again": lifestyle monitoring technology enables caregivers to see homecare clients in new ways. I will analyze here how processes of technologically mediated *seeing again* and of care's tinkering take shape in a specific compound in Dutch homecare, and how respect – or disregard – for clients as persons is part of emerging care practices. After an introduction to the system, I will focus on the three themes raised above: norms and individuality; risks and control; and care relationships.

2. LIFESTYLE MONITORING IN DUTCH HOMECARE

The Dutch project *Tailored Care through Monitoring Lifestyle* involves three homecare organizations in the South of the Netherlands an is financed by the general health insurance. Candidates for monitoring are selected by caregivers, but clients have to approve the actual installation. The system used, chosen because of its simple and affordable set-up, consists of five infra-red motion sensors that are placed in the bedroom, the living room, in and outside the bathroom, and in the fridge. The sensor data are sent to an external server, where they are interpreted for indications on the general level of activity in the home, bathroom use, nighttime activity and meal preparation. The system scans the data for acute and gradual changes in activity patterns and conclusions are displayed on a password-protected website, accessible to formal and informal caregivers. The first web pages provide an overview that links on to various specific tables with sensor data per room and point in time. In the overview, significant and sudden changes are marked by red dots, gradual changes by yellow dots, while green indicates activity within the normal range.

The system generates two kinds of alerts. In the Dutch project, sudden events that may indicate acute problems or falls – such as not getting up in the morning beyond a set time or staying in the bathroom for longer than an hour – are relayed through existing alarm systems. Yellow alerts are sent to the PDA of the coordinating caregiver for this particular client, who judges the alert and follows it up in her regular – sometimes daily – visits with the client. Extra sensors can be placed to detect nighttime wandering.

The ensuing analysis is based on observations of regular team meetings spread over an eight-month period, in which six teams of caregivers received training into the use of the system and discussed experiences and emerging dilemmas. I spoke with managers and trainers of similar projects in the Netherlands and Belgium and conducted in depth interviews with five care workers and one family caregiver. Two interviewed care workers also monitored a family member. During the interviews I discussed their experiences with caregivers, but also observed how they use the system's data screens. Personal names of caregivers have been altered for reasons of privacy.

3. Ruling Norms and Individual Rhythms

Lifestyle monitoring technologies make behavior readable in terms that are built into the system. A central set of these are the relevant activities measured – sleeping, eating, toileting and general activity. This is a stripped down version of the scales of Activities of Daily Living (ADL's), current in healthcare practice to score the degree in which clients are able to take care of themselves and assess disabilities. When thinking about personhood, we normally focus on aspects of our being that mark us off as individuals from other people, considering daily care for our self and our affairs as a self-evident and, indeed, largely shared baseline or norm. With aging, however, the taken for granted daily care can take on new significance for personhood. In a society that emphasizes autonomy increased dependence in basic activities is often experienced as a threat to personal dignity. The association of aging with growing incompetence is also a source of ageism in society at large. Not surprisingly, therefore, homecare clients are ambivalent about monitoring: while some welcome the feeling of being watched over, others fear the stigmatizing effect of the system. "They don't have to watch me from every corner," the father in law of Helen, a caregiver I interviewed, stated. "Those things are for old people," the mother of Natalie, another caregiver, declared. Clients often accept monitoring only after being urged by caregivers and family or, as was the case with Helen's father in law and Natalie's mother, when prompted by distressing accidents.

Basic norms about what it is to be a healthy, adult human being – one who washes, toilets, eats and sleeps according to regular patterns – structure what the system observes. Disturbance of regular eating patterns or activity at the wrong time in the wrong place, such as wandering in the night or sleeping in the living room, are specifically considered as symptoms of advancing dementia. By providing indications about these, critics suggest, monitoring invites observing clients according to deviations from 'normal' patterns.

In homecare, however, norms also have practical significance. Dutch homecare deals with personal hygiene, health and – to a lesser extent – emotional wellbeing. Disruptions in eating, toileting and sleeping typically signal a number of recurrent homecare issues, such as problems with blood sugar levels or metabolism. Being up in the night can point to adverse effects of administered medicine or pain. From the perspective of care, therefore, detecting irregularities is not simply framing a person as deviant. Symptoms appear as potential personal and practical problems. Restlessness in the night is often accompanied by anxiety and loneliness: Helen recalled the distress for the family and her father in law personally when he landed up in his nightclothes on the street several times within a matter of weeks. "One time he had to wait for morning to get back into the house," she said. The sensor installed by the door after these events gave everyone the reassurance to go to sleep, knowing there was a safety net in place.

Monitoring does not just detect adherence to norms, it also brings to light individuality within generality. In the first weeks following the installation of the sensors, the system establishes activity patterns belonging to this person, on which it bases its alerts. This opens the possibility for homecare to take individual living patterns of clients into account. For instance, when the system indicated that a client consistently rose hours before caregivers would come to wash and dress her, they were shocked and morning care was brought forward. An interviewed project manager pointed out that the attunement of existing work patterns of caregivers and organizations to those of clients is a new and still unsolved challenge to homecare. Technologically mediated seeing, here, confronted homecare with its former blindness and led to new, personalized obligations felt towards clients.

How can we conceive of the person that emerges through the mediation of lifestyle monitoring? Drawing on the work of Merleau-Ponty, Kontos (2005) points to aspects of selfhood that are located in the intentionality of the pre-reflective body, such as small, often taken for granted, gestures she observed in people with dementia or continued expressions of culturally acquired behavior that disclose "coherence and unity in their directedness towards the world" (p. 561). The self care registered by monitoring also involves much ingrained behavior that forms part of who we are as embodied and cultural beings on a pre-reflective level. Monitoring brings out typical routines in getting up and going to bed, and in the order, duration and frequency of washing, resting and meal preparation. These routines are interwoven with the immediate environment. As Rowles (2000) argues: "Over the duration of our lives, we each develop a rhythm and a routine in our use of space and in our relationships with the places of our lives that provide a sense of being in place" (52). Rhythms turn living places into parts of our personality and afford continuity in a changing world. As such, they gain special significance for frail people. Caregivers mentioned being struck by the observed consistency of patterns displayed by their clients. Habits, in this view, are not dull conformity to norms, but an expression of being able to live in-the-world and a vital part of our embodied identity.

Lifestyle monitoring thus operates in a field of tension between the inherited and normative and individual being-in-place. Seeing rhythms connects the quantitative where and when, detected by sensors and algorithms, with qualitative aspects of bodies living in space. Detecting rhythms is not computing averages; it requires observers to open their body to the resonance of emerging patterns (cp. Lefebvre 2004: 20-25). Within the monitoring compound, the observing body open to rhythms – as will become clearer below – is a composite of technology and the sensibility of human caregivers.

Nevertheless, the technological system as such misses much of the content of rhythmic activity. It detects how often and when a person opens the fridge, as an indication of meal preparation, but does not show that someone actually eats nor whether meals are made with relish or merely a bleak sense of obligation. The system may indicate that someone tends to sit in the living room after care visits, but not whether a client is watching television or telephoning friends, nor in what mood this is done. Its registration of being-in-place, therefore, is "reduced" (Ihde 1990: 88) when compared to observations made by people with whom one shares life intimately. Being used primarily for people who live alone, technological seeing partly remedies situations where no humans make those observations on a continuous basis.

Studying the routines of elderly people in the village of Colton, Rowles (2000) describes the identities of single people as a system in homeostasis, linked in with the material environment and the routines of other people. The lifestyle monitoring compound can be seen as such a system, detecting a given balance and aiming to remedy disruptions. Balance, much more than adherence to norms, describes what caregivers seek for their clients. "I don't get it how other people live [...] But I don't have to get it all [...]," says Brenda, another caregiver, "it's nice if there's a pattern." Still, as Lisa, another caregiver explains, "clients are different. This pattern may be wrong for another client."

4. Risks and Control: Seeing too little and Seeing too much

In western societies, "where risk management has burgeoned" (Lyon 2001: 6), detecting risks is an impetus for preventive intervention. Technological risk detection may turn the care for vulnerable people, Kenner (2008: 261) warns, into behavior control. But caring for frail people also demands attention to risks and, at times, taking over for persons who no longer manage on their own. Respecting personal space cannot be a license for neglect. Again lifestyle monitoring is operating in a field of tension, this time between control and care.

Envisioning risks may invite action, but also breeds anxiety; in the care for frail elderly people, the line between seeing or merely imagining risks is fuzzy. Monitoring is a technical response to the uncertainty, experienced by caregivers, about the safety and wellbeing of people in their care. It offers caregivers, in the words of a trainer, "extra eyes".

Over the past decades, budget cuts in homecare in the Netherlands, as elsewhere, have led to tight working schedules (cp. Kunneman/Slob 2007: 16). Caregivers have *hunches* about the wellbeing of clients, but limited opportunities to verify these. Modern mobility and labor patterns lead to situations where family caregivers have to divide their time between work, their own children and care for aging parents, and they do not always live close by. At the same time, frail elderly people remain living in their own home longer. In the Netherlands, over 40% of the people with dementia who still live at home, live alone (*NRC*, 05 October, 2013). Lifestyle monitoring targets the diffuse turning points where clients may need more or different care and where aging parents become increasingly dependent. These turning points are also difficult to determine because changes are often gradual and hard to pin down in one single factor. Natalie, a family caregiver with a job in healthcare and three children at home, relates her worries about her mother as follows:

Is she getting up at a regular time? Is she still eating? Is she actually still doing that? My brother does her shopping and we see it's less and less [...] She is very independent. She doesn't need anybody. Still, you notice in small things that she is slowly going downhill [...] What I hope from [monitoring] is to get a picture [...] She does all kinds of things, she says, but is that right? Craftwork is getting difficult. She is getting rid of a lot of stuff. From these things you notice, she is going down [...] I want to see whether anything is needed. And I want to be in time [...] Are things really alright?

For elderly people themselves the changes are gradual too. Moreover, many are ashamed to admit that they can no longer manage certain tasks. They do not want to be a burden and also fear the intrusion into their lives that comes with care. Still, many do have concerns about personal safety, such as having an accident and being unable to call for help.

Monitoring reshapes this situation by registering a defined set of activities, generating short and longer term overviews of activity patterns and alerts. For caregivers the system's observations provide orientation and set limits to diffuse and multiple worries. For clients it can provide the feeling they are being watched over. However, telecare systems do not "fit" all clients (Pols 2012: 40); in several cases, the system was removed. As Natalie quoted one client: "the flickering [of the sensor's control lights] drives me crazy …and then I have an alert and everyone asks me where I was and that is completely unnecessary." Brenda recalled a client that welcomed the sensors: "she really had contact with the sensors […] when going to the bathroom she would say 'Hi, here I am' […] each night when going to bed, she would say: 'okay, you can stop now I'm off to sleep' […] She felt 'I'm not alone now. There is someone there'." Being seen as "at risk" thus does not mean the same to every client.

The first screens caregivers see when accessing their clients' web pages offer summaries of recent activities in color code. These play an important role in guiding caregivers. "I check the color page daily, after work," Joan, a family caregiver whose mother has dementia, recounts: "is everything green, then it's okay. With yellow, I call." Her mother tends to forget to eat and yellow dots are often connected to meal preparation. In those cases, she calls her mother and walks her through making a meal. Mostly though, all dots are green. In general, the system reassures her that, despite her mothers' advancing dementia, the situation is still tenable. Anxiety of family is a main deciding factor for a move into residential care and caregivers mention reassurance as a key effect of the system that can suspend that move.

Professional caregivers also experience reassurance. Because they receive yellow alerts on their PDA, they do not need to check the website to know everything is green. "I always sort of wait for the alert in the morning… If I don't hear anything by 9 [the time that the system sends the yellow alerts generated over the previous day], then I know it's okay," says Helen. When the situation of their client is stable, caregivers partly rely on alerts to tell them whether they have to look into something, though most scan the overviews at least once a week.[1]

---

1 | Results from a questionnaire distributed among caregivers.

In sum, though monitoring structurally emphasizes risk, in practice the reassurance of green plays as significant a role as red and yellow alerts. "It is hard for us to work with our hands behind our backs," Norma explained, "but sometimes you provide better care by not doing anything." Monitoring supports intervention, but also drawing back. Price (2009) confirms this effect for a comparable system used in England and even speaks of people with dementia being given "new means to communicate their capabilities" (13).

Whether it communicates unexpected abilities or inabilities or confirmations of the already known, the *extra eyes* that the system provides are taken by caregivers as seeing better than human eyes in some respects. They keep watching when caregivers leave, are supposedly more objective, and better equipped to detect gradual change. On this basis the system reassures, warns or confirms. Interestingly, though caregivers welcome their extended eyes, they also draw their own boundaries about what they want to see. Green also signifies there is no need to check clients' data. Privacy concerns figure prominently among issues caregivers raised and they are sensitive to their clients' personal space: "It is not like I have to know people through and through," Natalie stated, "For me it doesn't matter whether I can see that someone gets up at 6 or 6.30. That is no extra information for me." And Lisa explained: "You know, it's not like we are sitting down and start to look at everything." The rule she formulated, "you should not check more than you need," aptly articulates the attitude caregivers espoused in meetings and interviews.

Caregivers feel awkward about tables that give very specific information. "I don't have to know where she is," Brenda stated about tables that show her client's activity per room in the last 24 hours, "I don't like it. I want to know she is home, not more [...] I just want something I can do something with." What information is useful differs from client to client, though. Brenda also recalled a situation where it was helpful to see that a client, who had severe dementia, stopped using her bedroom: "She was getting very frail and couldn't do it, so we moved her bed into the living room." Working with compound eyes makes caregivers see better, but also compels them to negotiate when and where to look and when to close their eyes in order not to infringe on their clients' personal space.

Technological monitoring does affect power relations inherent in care: caregivers see things their clients cannot or do not want to tell them while clients' verbal accounts are bypassed by sensor data. When the clients' cognition or call for help fails, the system can stand in for them and make their technologically enhanced bodies speak and allow needs to become visible. It can also betray what they want to hide.

Clients are not merely passive users. As Haraway (2008) reminds us, relationships within techno-cultural compounds are not necessarily symmetrical, but all composites are "at work" (262). Once the sensors are installed, only the clients' co-operation ensures meaningful reading of their bodies. Clients have to refrain from meddling with the sensors, they have to entrust themselves to the system and the observation of caregivers, and allow these to bring up issues. The system asks them to be, at the same time, generally aware of the security provided, but to forget its presence on a daily basis. Then their extended bodies can express their being-in-place to caregivers beyond the scope of immediate proximity.

As such, monitoring can relieve them from having to tell something is wrong. As Natalie stated about her fiercely independent mother:

I hope that when something is wrong, she doesn't have to cross that threshold: now I have to call her that I cannot manage anymore. That I can simply say, Mum, I saw this or that. Wouldn't you want such or so? […] If through this system, we can go to her, than that threshold is already gone.

A professional care worker herself she explained clients find it hard to request care: monitoring opens the discussion of needs and circumvents threats to personal dignity. Lisa related how one client did not tell her about her persistent diarrhea but she noticed that her client went to the bathroom a lot and inquired about her disturbed sleep. She recounted how clients hesitate to mention shameful problems to various caregivers, which turns distressing situations into something that is simply accepted. Monitoring, she found, creates a context for understanding, "the space has already been opened."

Understanding needs is also significant in another respect. Caregivers don't always know how to interpret information their client gives them. People with dementia in particular may not remember whether they ate or slept well. A client declares she slept badly. Does this signify anxiety that needs to be met with sympathy or is there a physical problem? One of Helen's clients did complain about diarrhea and being up all night. But she was also the kind of person that was often low-spirited, seeking justifications for not having to go out. Was the diarrhea one of these? Monitoring confirmed her story and the problem was solved by adjusting medication.

The monitoring system does not decide how to intervene: its data serve as a reference point for negotiation by supposedly providing objective observations. Yet, subjectivity resurfaces in the interpretation and weighing of factors involved. Monitoring supports caregivers' interpretation of care that is needed in ambiguous situations, but care interventions also require other modes of seeing that confront the system's observations with singular clients and their particular situations. Disturbed sleep might be compensated by offering more activity during the day, disturbed eating patterns by inviting meal service, but with clients that appear "into their own" caregivers also disregard the system's information. One of Norma's clients is 95. She stays up a lot at night and eats like a bird:

[…] you can't change that. It is her pattern of living. I can't go there and send her to bed at 23.00h! She has always been a nighthawk. She doesn't do that much, she sits in her chair all day […] To her it is fine like that […] I sometimes get meal alerts. I know she eats badly. Soup and biscuits. She made it to 95 on those. It is not my role to meddle with her life pattern.

Deciding on interventions is a precarious balancing act. Helen recalled a client with dementia who suffered from diabetes:

She couldn't explain to me how she was doing with eating, or in the night […] She had very irregular sugar levels. She would just eat a whole pie. [With monitoring] I could see when she used the fridge and how it related to the sugar levels […] I don't need to know that she eats three bowls of ice cream, but if her sugar is high, it is good to know[…] that you then don't adjust the insulin.

The technological eyes helped Helen to steer her intervention with regard to medication, but they didn't solve the discomfort she felt about how far to go with stopping her client's unhealthy eating habits. Care involves weighing the wishes, personality and abilities of clients against what is good or feasible. These issues are usually discussed with clients or their family. Monitoring, it turned out, stimulated the conversation with family because it provides a shared information platform (cp. Willems et al. 2011: 177).

Good care also requires that caregivers ask what it means that a client gives different information than what is seen through the eyes of technology: "I don't press on", one caregiver said, "she doesn't have to let me in on everything." Respecting someone's privacy, however, can be wrong. Lisa mentioned a client that used to remain in bed during care visits, seemingly incapable of handling her morning chores: monitoring revealed that she was quite active after care visits. Lisa concluded that her client's need was less to have meals prepared for her, and more to receive some tender attention. She decided to prepare breakfast together with her, which combined activity with socializing.

Discrepancies in clients' accounts and data displays can thus lead to reinterpreting needs. To acknowledge this as part of care as "person work" (Kitwood 1997: 69), we have to give up the idea that clients unequivocally know and voice what they need. As Mol (2008: 11-12) argues, healthcare clients are not consumers, buying products, nor citizens, claiming their rights. Care is a cooperative and ongoing search for what works in each specific situation for each individual person. Where discourses of rights dominate general debates on surveillance, pragmatics and a focus on clients as individual persons are better guidelines for care's tinkering.

5. CARE RELATIONSHIPS

Motion sensors do not register individual activity and only deliver usable data for people who live alone. Therefore, monitoring could increase the isolation of frail elderly people when it replaces actual care visits—the last is indeed a benefit expected by health insurance. Two factors limit this in practice. Firstly, only visits that do not involve physical action can be reduced. The vulnerable clients for whom the system is used receive between two to six visits a day for support with washing, dressing, meal preparation, housecleaning and administering medicine.

A second factor is more complex. Its developers present the system as a tool to increase the "accuracy, reliability and validity" of care assessment (Glascock/Kutzik 2006: 60). They argue for "appropriate" care, based on "accurate knowledge" (p. 59). The language used is that of healthcare policy, the image drawn up one of seamlessly targeted care. Oleson (2006: 245-246) discusses the tendency in healthcare to move from multisensorial embodied perception toward explicit knowledge, based on hermeneutic readings of instruments. Instruments support healthcare as a specialized, professional discipline. As a result, the distance between caregivers and clients and between personal judgment and professional work is often enhanced.

True as this may ring in general, the term 'knowledge' in the case of monitoring is misleading. Without familiarity with a client's life and home situation, the dataset gained by monitoring is a lifeless corpse. A sudden rise in activity in the home can be due to family visiting, and a lack of fridge use to eating out. Caregivers know these things about their clients and use their background information to interpret data and weigh alerts. Monitoring non-homecare clients proved a failure. Alerts could not be put into perspective, which resulted in frequent telephone inquiries and mounting frustration with both clients and caregivers. Lifestyle monitoring's observation, it appears, works best as a *seeing again*, within the context of existing relationships.

"If you aren't around for a week or so, everything keeps going," Lisa explained why she has come to like the system, "you have something to talk about when you come back." To her, monitoring provides continuity in the care relationship. Caregivers indicated feeling closer to their clients. Embodied and hermeneutic forms of perception need each other here: together they increase the familiarity that is at the same time a prerequisite to work with the system. Caregivers' articulations of their experience do point to a difference between having hunches and a form of knowledge, but also indicate how this 'knowledge' is embedded in relationships, confirms and relieves worries, or can steer to someone's true needs.

Though caregivers speak of "seeing" and "getting a better picture" through monitoring, they don't just see, but also *hear* alerts. They do challenging work in interpreting data screens that demand different ways of perceiving, from reading symbolic code, to lists and tables to almost physically feeling activity in the graphs. Increasing familiarity, both with the system and their client, makes them recognize specific patterns as typical for their client. When discussing data displays, they often referred immediately to particular situations. Hermeneutic perception, with the help of contextual knowledge and imagination, thus turns into an embodied feeling of clients through the system. "This is her," exclaimed Norma, who found computers still daunting, during our interview and pointed to a blue line in a graph. While she relegated the rest of the frantic activity in the home to family visiting, this line, around 1 AM represented her client: "she always goes to bed late. Undressing takes her a long time. She washes carefully and lays out things meticulously, all by herself." The sight on the screen actually made her smile.

Often the monitoring system functions largely unnoticed, so occasions when the system becomes explicit pose challenges to the care relationship. One such moment is when caregivers make observations that need follow-up; for clients the confrontation with caregivers' inquiries means waking up to the presence of the system and when this happens without sufficient reason they are easily disturbed. Therefore, caregivers often don't discuss alerts that they judge meaningless: "The system must take the background," says Helen, "otherwise [clients] really are watched and that's what you don't want." With potential problems, however, caregivers have to confront clients and this can be risky for the care relationship. Some caregivers avoid this and refer to family or colleagues to detect what is going on. Norma's claim that the system is only suited to clients who are no longer aware of it mostly reflects her own discomfort with having to communicate about observations. Other caregivers are better able to fit their new role into their care. When she has to inquire with her client, Brenda always recounts why the system was installed and explains what she saw: "You have to be absolutely honest," she stated. Lisa sees no problem at all: monitoring is so much part of her care that she daily brings things up openly. Natalie related how, after a few weeks, monitoring became the object of jokes between her and her mother: monitoring had to be fitted in their relationship and trust in each other's new roles had to grow. Brenda stated that she felt grateful about her client's faith in her and that monitoring had brought them closer. In order to work well, monitoring needs to be embedded in relationships of trust; it can then make that relationship more explicit and communicative.

6. Conclusion

Seeing and appearing within the monitoring compound takes shape against a background of not wanting to or not being able to see or appear. Monitoring is not fit for every client, no matter how vulnerable they seem to caregivers. For clients who welcome it, monitoring can serve as an expression of abilities and needs that they cannot or do not articulate clearly. Ideally, care means being attentive and responsive. In practice, it also involves not seeing, due to work pressures and limitations in caregivers' perception and sensibility. Technologically enhanced seeing partly compensates their blindness and enables more familiarity with clients' individual rhythms and needs. To reach this goal, however, technologically enhanced *seeing again* requires physical familiarity. Moreover, caregivers also have to compensate the structural blindness of their extra eyes by re-considering technologically generated data in the light of a particular person and her context. Interpreting data and deciding about care require both *seeing with* and *seeing against* the technology, and at times, technology's seeing even compels human eyes to be averted to safeguard clients' personal space.

Monitoring functions within care as a relationship of trust. The system's design could use improvement. At present it quietly refers data out of the home, while little in its set-up communicates to clients about *their* data and whether these are in good hands. Securing trust now xfalls mostly to caregivers. In order to safeguard or even strengthen care relationships, clients and caregivers both have to learn to trust the system and each other. Fears that lifestyle monitoring is necessarily normalizing, disempowering or objectifying do not seem justified. Securing *respecere* in a multiple sense, however, does demand that technological eyes work in conjunction with human competences, such as particular perceptual and communication skills, sensibility and with values such as respect for persons, honesty and trust. Though part of good homecare, in the face of new technologies, these demand education and deliberation, while management and insurance have to secure the proper context for care, worthy of trust.

Acknowledgements

This research was made possible by financial support from ZonMW, Disability Studies, and the cooperation of Proteion Thuis (The Netherlands), for which I am very grateful. Zuyd University kindly gave access to provisionary case descriptions.

References

Baldwin C. (2005): "Technology, Dementia, and Ethics: Rethinking the Issues." In: Disability Studies Quarterly 25/3, http://dsq-sds.org/article/view/583/760. Retrieved: November 07, 2014.

Bowes, A./Dawson, A./Bell, D. (2012): "Ethical Implications of Lifestyle Monitoring Data in Ageing Research." In: Information, Communication & Society 15/1, pp. 5-22.

Brittain, K./Corner, L./Robinson, L./Bond, J. (2010): "Ageing in Place and Technologies of Place." In: Joyce, K./Loe, M. (eds.), Technogenarians: Studying Health and Illness Through an Age-ing, Science, and Technology Lens, Chichester: Wiley-Blackwell, pp. 97-111.

Coronato, A./De Pietro, G. (2010): Pervasive and Smart Technologies for Healthcare. Ubiquitous Methodologies and Tools, Hershey: IGI-Global.

Glascock, A.P./Kutzik, D.M. (2006): "The Impact of Behavioral Monitoring Technology on the Provision of Healthcare in the Home." In: Journal of Universal Computer Science 12, pp. 59-78.

Haraway, D.J. (2008): When Species Meet, Minneapolis: University of Minnesota Press.

Hughes, J.C./Louw, S.J./Sabat, S.R. (2006): Dementia. Mind, Meaning, and the Person, Oxford: Oxford University Press.

Ihde, D. (1990): Technology and the Lifeworld. From Garden to Earth, Bloomington: Indiana University Press.

Kenner, A.M. (2008): "Securing the Elderly Body: Dementia, Surveillance, and the Politics of 'Aging in Place'." In: Surveillance & Society 5, pp. 252-269.

Kitwood, T. (1997): Dementia Reconsidered: The Person Comes First, Buckingham: Open University Press.

Kontos, P.C. (2005): "Embodied Selfhood in Alzheimer's Disease: Rethinking Person-centred Care." In: Dementia: The International Journal of Social Research and Practice 4, pp.553-570.

Kunneman, H./Slob, M. (2007): Thuiszorg in Transitie. Research report, Bunnik.

Lefebvre, H. (2004 [1992]): Rhythmanalysis. Space, Time and Everyday Life, London: Continuum.

Lyon, D. (2001): Surveillance Society: Monitoring Everyday Life, Buckingham: Open University Press.

Mahoney, D.F./Purtilo R.B./Webbe, F.M./Alwan, M./Bharucha, A.J./Adlam, T.D./Jimison, H.B./Turner, B./Becker, A./for the Working Group on Technology of the Alzheimer's Association (2007): "In-home Monitoring of Persons with Dementia: Ethical Guidelines for Technology Research and Development." In: Alzheimer's &   Dementia 3, pp. 217-226.

Mol, A. (2008): The Logic of Care: Health and the Problem of Patient Choice, London: Routledge.

Ni Scanaill, C./Carew, S./Barralon, P./Noury, N. /Lyons, D. /Lyons, G.M. (2006): "A Review of Approaches to Mobility Telemonitoring of the Elderly in their Living Environment." In: Annals of Biomedical Engineering 34/4, pp. 547-563.

Oleson, F. (2006): "Technological Mediation and Embodied Health-care Practices." In: Selinger E, Postphenomenology: a Critical Companion to Ihde, Albany: State University of New York Press, pp. 231-247.

Pols, J. (2012): Care at a Distance. On the Closeness of Technology, Amsterdam: Amsterdam University Press.

Price, C. (2009): "Just Checking: Lessons Three Years On." In: Journal of Dementia Care 17/3, pp. 12-13.

Rowles, G.D. (2000): "Habituation and Being in Place." In: The Occupational Therapy Journal of Research, 20S, pp. 52-67.

"Tachtigduizend dementerenden wonen thuis alleen", October 05, 2013 (http://www.nrc.nl/nieuws/2013/10/05/tachtigduizend-dementerenden-wonen-thuis-alleen/). Retrieved: June 23, 2015.

Willems, C.G./Spreeuwenberg, M.D./van der Heide, L.A./de Witte, L.P./Rietman, J. (2011): "The Introduction of Activity Monitoring as Part of Care Delivery to Independently Living Seniors." In: Bos, L./Dumay, A./Goldschhmidt, L./Verhenneman, G./Yogesan, K. (eds.), Handbook of Digital Homecare: Successes and Failures, Berlin: Springer, pp. 167-179.

Enabling a Mobile and Independent Way of Life for People with Dementia – Needs-oriented Technology Development

NORA WEINBERGER, BETTINA-JOHANNA KRINGS, MICHAEL DECKER

> "I run around in the corridors of my memory and feverishly trying to understand what's going on. Sometimes the search makes me even more confused, as I forget what confuses me."
> TAYLOR 2008: 6

1. INTRODUCTION

Life expectancy will increase over the next 50 years, and demographic change over the coming decades appears to alter Germany's population structure. These developments offer the opportunity to integrate older people in their neighbourhood[1] as long as possible, but they also raise questions around how people are cared for and looked after in the future. In 2011 2.5 million people (83% aged 65 or over) were in need of care in Germany; 754,000 people were cared for in 12,400 licensed nursing homes (Federal Statistical Office 2015). Furthermore, given the rapid growth of people suffering from dementia (Bachman et al. 1993) and the parallel decrease in the number of people entering the care profession, the unmet needs of people with dementia will grow to an even greater extent (Meiland et al. 2010: 80). In Germany, more than 1.4 million people suffer from moderate to severe dementia (German Alzheimer Society 2014); two-thirds of them are affected

---
1 | The term 'neighbourhood' denotes here and hereafter a defined, public and social space around a residential care facility.

by Alzheimer's. Every year nearly 300,000 new cases occur. Unless there is a breakthrough in prevention and therapy the number of people with dementia will increase to about 3 million by 2050, according to projections of population development. This corresponds to an average increase in the number of sufferers of around 40,000 per year, or more than 100 per day (ibid). It is already assumed that 69% of residents in nursing homes suffer from dementia-related mental illnesses (BMFSFJ 2007). Based on these demographic and social developments shortages in and pressure on the provision of care are expected. Affordable and high quality care must be guaranteed for the expanding population of those who need it.

An increasing number of scholars propose technological solutions like assisted living strategies to address this challenge, as they have the potential to enhance the efficiency and effectiveness of formal health care services. These technologies could improve and extend the quality of life for people with dementia by helping them lead fuller and more independent lives. In this context, various types of services and several devices are discussed and are available (see, inter alia, Lenker et al. 2013; Salminen et al. 2009; Löfqvist et al. 2005). Currently there are relatively simple products on the market such as automatic pill dispensers, as well as more complex and complete products such as tracking devices using Global Positioning Systems (GPS) that assist in locating people. In addition, internet-based applications designed to provide carers with clinical, decision-making, and emotional support have been evaluated in field trials and the initial results have shown the systems to be beneficial both to carers and to people with dementia (Lauriks et al. 2007). Nonetheless, at present there are just a limited number of systems on the market which are capable of offering help and solutions with respect to movement for people with dementia in long-term care. This could be because the movement needs of people with dementia are very different. In general, it must be noted that currently available applications or devices for people with dementia have been designed with little involvement of the end-user (see in this regard Encarnação, et al. 2013; Orpwood et al. 2005). Where there was involvement of the end-user, this occurred after the decision had already been made as to which kind of device or service would go forward for technology development. It is our view that integrating and respecting the perspectives of the whole care arrangement[2] (people with dementia, their

---

2 | The authors support the use of the term 'care arrangement' (Blinkert und Klie 2004) which has been further developed by their organisation, the Institute for

relatives and formal carers) in the development process has a decisive impact on the development of technologies for care and support to become successful innovations. Technologies are developed for specific, everyday situations, according to the social context of the technology and the living environment of the potential user.

2. MOVEMENT IN PEOPLE WITH DEMENTIA

Dementia is an umbrella term used for one of the most common mental health problems amongst elderly people; it is progressive and characterised by serious losses in cognitive functioning, particularly memory, thought processes, orientation, aptitude for learning, language, as well as the ability to judge. Along with these cognitive losses, emotional and behavioural changes occur, which can develop into depression, hallucinations and extreme agitation (Schäufele 2008: 169). As the dementia progresses so does the loss of mental ability, including the capacity to solve everyday problems, with the effect that people with dementia experience an increasing loss of independence and rely ever more on support. This deterioration can express itself in a great variety of ways and differs greatly between individuals. A relatively common form of expression is extreme agitation amongst people with dementia, which can manifest itself as a strong urge to move (known as 'wandering'). This, in combination with the impairments described, can lead to elderly people potentially endangering themselves, as their orientation ability is limited or non-existent. In addition, people with dementia displaying these symptoms may walk themselves to exhaustion, and fail to adequately perceive their own physical limitations or bodily signals, such as hunger or thirst. Other people with dementia display so-called 'running away tendencies', leaving their neighbourhood and wandering in search of previous environs (a house they lived in, perhaps), or wander because they sense the need to complete a chore (look after their mother, for example).

At the same time, movement is recommended as an intervention to activate brain function and thereby slow down the development of symptoms, and

---

Technology Assessment and Systems Analysis at the Karlsruher Institute of Technology (KIT) (Krings et al. 2013). The term captures the relationship between care needs and care activities. The care arrangement concept analyses how and to what extent various actors are involved in shaping care in real situations.

is recognised as countering gradual immobilisation (becoming bed-bound) (Schrank 2013; Abt-Zegelin/Reuther 2011; World Health Organization 2010; Zegelin 2010). The authors of the S3-Guidelines 'Dementia' by the German Society for Psychiatry and Neurology (Deutsche Gesellschaft für Psychiatrie, Psychotherapie und Nervenheilkunde) and the German Association of Neurology (Deutsche Gesellschaft für Neurologie) reach the following conclusions regarding movement facilitation as a psychosocial intervention: "Regular physical movement and an active mental and social life should be recommended" (DGPPN/DGN 2013: 91). The authors refer here to studies which evaluate an active lifestyle with physical movement, sporting, social and mental activity as protective with regards to the development of dementia (ibid: 91). Another supporting argument is that falls are one of the major health risks in our rapidly ageing population. Falls frequently result in moderate to severe injuries and when added to the fear of falling, they can limit the activity of the people with dementia (Baldewijns et al. 2013); the mobility and balance of the person already at risk thus further declines, subsequently increasing the risk of future falls (Fleming/Brayne 2008; Milisen et al. 2004).

Movement can therefore be considered an effective and supplementary key component in the care and support of people with dementia. Movement is associated with motor, sensory and social activation, which can have an effect on the quality of life and functional status of people with dementia, and can help avoid contractures and bed sores. In addition, the development of dementia in older people often occurs in association with signs of ageing, and according to Kangas et al. (2009), about one-third of all people over the age of 65 fall once a year. In many residential care facilities efforts are made to reduce the number of falls by strengthening 'body resources', as there is so much scientific evidence that exercise itself is important for the body's physical well-being, but no "vision-zero concept" will ever be reached in relation to falls. Movement, physical activation and inspiring physical strength also fulfil basic psychological and emotional needs. Individuals' own capacity to move may be tested and developed autonomously. Through the experience of this capacity to move, the daily routine can be designed to be vital and stimulating in order to maintain existing capabilities for as long as possible and thereby delay the need for a high level of care.

Enabling movement, in particular outside the care facility, poses a difficult situation for care providers, residents and their relatives, as well as for the inhabitants of the neighbourhood: on the one hand, independence for people with dementia is to be sought and promoted, whilst on the other, depending on the individual's cognitive abilities and their condition, carers and relatives seek safety for the elderly person, which often correlates with a withdrawal of freedoms. This leads to a situation in which the desired autonomy and independence is restricted by safety considerations and a fear of self-endangerment. All of these issues considered, the care facility itself often has regulations to ensure its duty of care, which don't allow for movement in the open air. As a result, legal reasons dictate that the elderly person's safety is given a greater weight than their entitlement to movement. The frequent shortage of sufficient personnel in these facilities often contributes to the fact that individually-designed movement facilities cannot be provided.

As a result of the situation in residential care a number of conflict areas can be identified, namely the aforementioned conflict between freedom to move and self-determination versus safety.

Upon this backdrop, the question is whether and how technological devices yet-to-be-developed can support (in various ways) individuals' ability to participate and carry out daily life as independently as possible, and thereby maintain and develop the basic need for movement within care contexts (see, inter alia, Skymne et al. 2012; Pape et al. 2002; Daley/Spinks 2000). This question includes the intention to incorporate danger-free access to individual social areas in the neighbourhood outside of the care facility. This means that, using technically-possible devices, models are to be investigated which offer people with dementia the opportunity to (once again) take walks unaccompanied, go shopping or make visits within the social neighbourhood in which their care facility is embedded. In order to achieve this, innovations which strive to combine technical solutions with social interventions in neighbourhoods should be sought. Finally, steps should be taken so that people with dementia, as members of the neighbourhood, encounter a dementia-friendly environment and an age-friendly infrastructure. In this situation, technical aids could provide a reliable basis upon which people with dementia are able once again to manage the external world. Moreover, other technical systems which offer elderly people numerous options for reliable orientation are conceivable: these could help reduce feelings of fear, doubt or lack of orientation and, in the short-term, help encourage steps out into the neighbourhood. In this context, technical options should be initiated and launched which enable the basic requirement for movement to be implemented in a range of management strategies. In the same way that there are great differences in the ability to learn new technologies during one's lifetime, the necessary adaptations and systems transferability in technical implementation will reflect changing abilities as the dementia progresses. Whether or not this will be successful, and the requirements potential users in care management will have of the technologies, must be verified in a demand-oriented analysis.

3. INVESTIGATING THE NEED FOR MOBILITY: THE IMPERATIVE OF NEEDS-ORIENTATION

Mobility plays a pivotal role in modern Western societies, and a survey revealed it to be the impairment-related issue of greatest personal importance (Roentgen et al. 2013: 571). For this reason, the motto for the care and support of people with dementia must be 'As much freedom as possible, whilst providing as much protection as possible'. As discussed above and according to experts, technical support to help in this area of conflict suggests itself and is already being developed, e.g. localisation devices. Despite the fact that various scientific disciplines emphasise the importance of user-oriented development of products and services and that according to surveys manufacturers rank usability as 'high' or 'very high' in relation to market success, in practice implementation is typically technology-oriented. This means the integration of a potential user occurs in order to optimise a technology's usability. However, these development processes remain technology-focused, as they only take into account the needs of the potential user in relation to an already specified technology (cf. Cooper 2002; Reichwald et al. 2004; Bias/ Mayhew 2005). As a result, only traditional market research was carried out until the start of the buying process, or during product utilisation (e.g. complaint management). User tests are only carried out once the product is almost completely developed and, therefore, only minor modifications remain possible. With this 'technology-push' approach, a technology is suggested, which, from a technical perspective, fulfils care and therapeutic needs. This is particularly relevant for assistive technologies and services, given that these are still very new areas of research (Nygard/Starkhammar 2007).

Reasons for this are, on the one hand, a lack of anthropometric data or human models for the very heterogeneous target group of elderly people with dementia and, on the other, the lack of life experience and the knowledge gap e.g. for gerontological aspects and specific user demands in many product developers. This shortcoming hinders the developers' ability to put themselves in the position of elderly people, in particular those with dementia, and stops them relating empathetically to their requirements. In this case, knowledge of individual disease symptoms and progression would be crucial in the development of user-oriented technology. In order that assistive technologies are successful on the market and aren't obstructed by use-related innovation issues, they must be acceptable to the user, that is, products must be user-friendly (i.e. fit for purpose) (e.g. Grunwald 2011; Schlick et al. 2010; Glende et al. 2009; Blythe et al. 2005). Moreover, the solution potential fails more profoundly the more specifically the relevant context is analysed. Increasing numbers of studies indicate that all 'users', (i.e. those with dementia), the care providers (several aspects of the changes in care work are focused on) and the relatives of those in care should be included (Weinberger et al. 2014).

4. IMPLEMENTATION OF A NEEDS ANALYSIS: THE MOVEMENZ PROJECT

Against the observations and reflections mentioned above, the project "Mobile, self-determined living for people with dementia in neighbourhoods" (Movemenz), funded by the German Federal Ministry of Research and Development (BMBF; 01/2013-12/2015), currently explores technical requirements – expressed as needs by potential users – for mobile technologies, primarily to provide the infrastructural preconditions and conceptual standards to allow people with dementia to move as autonomously as possible around their neighbourhood. Taking this into consideration, the course of this project follows a needs-oriented technology development and thus, analyses the 'desires' of potential users for technical support. Potential user, in this context, means the respective stakeholders in the care process, such as dementia patients, relatives, care professionals and service providers (the so-called care arrangement defined in Krings et al. 2013).

In particular, using a participatory design[3] of technologies, those with dementia are considered experts. The technological solution will not itself be assessed, but rather it will be regarded as just one component, with its on-going interdependencies with the stakeholders in the complete care arran-

---

3 | Participatory Design is the name given to certain design principles and practices which aim to create products and systems that are more receptive to human needs (Clement/Van den Besselaar 2004). Sanders (2002) differentiates it from the more commonly used term "user-centered design" (UCD) by describing UCD as design for users and PD as design with users.

gement. In addition, problems, expectations, feelings of security and accessibility (Gimskär/Hjalmarson 2013) or failings in the status quo (in short, the participating stakeholders' wishes) are to be included for consideration. Furthermore, these wishes must be explained to all stakeholders, potential conflicting wishes discussed and the expectations of the parties involved in the technology usage must be reconciled. In parallel, a methodological overview of subjective parameters, such as well-being and happiness, security, privacy as well as a clarification of the meaning of autonomy and quality of life will be undertaken. Quality of life implies, here, the involvement of the physical and social environment, quality of support, behaviour, medical and cognitive status, psychopathology and behavioural traits as well as experiences and emotional sensitivity in support providers. Drawing attention on these issues it becomes essential to identify the environmental, social and functionality contexts in which the technology is to be introduced and to analyse the socio-structural background of the care 'environment' in which the investigation is to be carried out. This approach implements the idea of social innovation[4]. Here the technical innovation is analysed in its use context. In this case, the use context of the technological development consists of the care arrangement and the neighbourhood. This is where the stakeholders' expectations as well as their 'unrealistic demands' meet: People with dementia may not recognise their failing capacities and will expect their relatives to allow them to live an autonomous life. Relatives form expectations of the people with dementia and the care personnel, whom they expect to be concerned with the security of the dementia sufferer and provide stimulating care for as long as possible. Finally, the care personnel expect support from the relatives and have a realistic expectation of the intensity of care and support.

---

4 | Up until now scientific and political interest was very much concentrated on technical innovation. Social innovation is a complex process that profoundly changes the basic routines, resource and authority flows, or the beliefs of the social system in which it occurs (Wesley/Antadze 2010: 2; Blättel-Mink 2006). Social innovation does not necessarily involve a commercial interest, though it does not preclude such interest. More definitively, social innovation is oriented towards making a change at the systemic level. As Phills et al. (2008: 37) explain, "Social innovation transcends sectors, levels of analysis, and methods to discover the processes – the strategies, tactics, and theories of change – that produce lasting impact."

In addition, social changes which accompany each technical innovation must be considered, thereby increasing the chances of the technical innovation's successful implementation, based on its acceptability. This means that all stakeholders in the care arrangement will be studied and assessed during the context-specific needs analysis. The user needs analysis is especially important, with a focus on the mobility of people with dementia, as effects like cognitive difficulties, physical disabilities and changes in emotions and behaviour vary from person to person. Therefore, the issues around people with dementia being mobile are person-specific. One person may need just a little balancing support whilst standing, while another may need intensive support or a technical device e.g. a walking stick, four-wheel walker or wheelchair (simple, low-tech). Not only does the need for support vary greatly between users, but technology offering similar functionality may have different actual uses. In addition to this, each individual's technological competence as learned over a lifetime must be taken into account as well as the continuous decrease of this competence in progress of dementia.

The methodological implementation of the user needs assessment should therefore clarify the conditions under which the support system would be attractive to all participating stakeholders. The starting point for the research should be an open needs assessment.

Starting with the identification of the empirical setting, determining the regional location of the home for people with dementia and gaining access to the field was challenging. The same applies to selecting which scientific methodology to use to provide an insight into the daily routines both of the people with dementia and all "care givers" (in the broad sense) around them. Coming from this perspective it seemed important to accompany and to simply observe the daily routines in the home. This included analysis of the institutionalised structure of the home on the one side, and on the other, it included an observation of how walks and shopping trips by the elderly people are conducted in long-term care settings, the environment in which their everyday life is embedded and which areas of need can be identified. Observation of technical aids within these routines was also a focus of attention.

Taking the challenges of a demand-analysis into account, the approach to the empirical field was developed into a two-phase process. In the first phase, sensitive involvement of the research team in the home as well as careful observation of the daily lives of the people with dementia was prioritised. In the second phase, individual as well as group interviews were conducted in order to hear the voices and opinions of *all* social groups involved.

In addition, the needs assessment is accompanied by an interdisciplinary and transdisciplinary research process. In this reflection process aspects of ethics, law, economy, technology, society and care science are observed based on empirical analysis. The aim is to consider not only technical but all relevant aspects, especially ethical and social ones but also, for example, aspects like who is liable for damages that might occur during the use of technical aids and what is the cost-benefit ratio of technological and social innovations.[5]

5. Approach to the Empirical Field and Initial Results

In order to get closer to the field of residential care of people with dementia and to learn something about health care and mobility relating to dementia in everyday care, the field was firstly considered from the perspective of the qualitative method of Grounded Theory.[6]

According to the ‚all is data' concept (Glaser in: Przyborski/Wohlrab-Sahr 2010: 198), materials and information such as the website of the residence, documents, statistics and informal talks etc. were included in the process of collecting data. However, explicit preference is given to observation, which is reflected within the observation protocols of each team member. These protocols are based on two two-week observation periods. This stage investigated how care is organised in these contexts, and which needs and challenges those participating have around the concept of movement. This specific method highlights that the 'field', being an essential pre-condition, is to be determined in an unbiased way. Unbiased here means that no technical options are presented which would affect the structure of the 'field' in advance. Instead, first and foremost, the intention is to open up the 'field of residential care' as the basis

---

[5] The entire project process is described elsewhere (see inter alia Weinberger et al. 2014).

[6] The process of Grounded Theory was developed by Anselm Strauss and Barney Glaser in the USA in the 1950's and 1960's. The basis of this qualitative applied method is to bring empirical research and theory-building closer together from the start. No specific surveys were developed within the scope of this method: "Important, here, is the interwoven process of sampling and theory generation according to the principle of theoretical sampling." (Przyborski/Wohlrab-Sahr 2010: 189)

of this exploratory project, as well as for general further questioning. Based on this, this phase was conducted in a care home in a neighbourhood with approximately 5,000 inhabitants in Baden-Wuerttemberg. The denominational, diocesan-supported, non-profit home for people with dementia is located in the redeveloped town centre, surrounded by historical buildings and shops of all kinds. "You can linger in cafes, or buy daily provisions in various specialist shops or markets. Churches, the town hall and the station are all easily accessible on foot" (care home's website, translated by authors). In this care home live 49 people aged between 76 and 98 (and one 66-year-old) spread over two floors (24 residents on first floor, 25 on the second floor). The residents suffer from moderate to severe dementia.

Based on these comprehensive observations and impressions, structured interviews were conducted, as mentioned above, to prioritise the areas of need according to the dementia sufferer's abilities. Targeted exploratory focus groups of relatives, carers as well as inhabitants of the neighbourhood were established in order to reasonably determine their motivations, expectations and fears. Not before the end of the process of needs assessment, questions were posed to the potential users regarding technical solutions with regard to possible technologies in the immediate future. The questions didn't focus, at this stage, on specific ideas for technological options. Instead, the questions were kept open e.g.: In what areas of care and support for people with dementia are technical aids needed for supporting everyday care?' and ‚what demands would be made of these technologies and how might they look?'.

Although the results of these two phases of data collection have not yet been systematically evaluated, the following chapter will provide some insights, as preliminary results. This preview will offer an idea as to how multi-faceted the implementation of a needs-analysis could be. This is particularly the case when technical options are adjusted to the different social needs of dementia sufferers' care.

6. Hypotheses based on Observation of the Empirical Field

As mentioned above, the data resulting from implementing Grounded Theory as a methodological approach, including observations protocols, discussion notes, photographic material as well as complete interview materials, was systematically evaluated. In the first stage of analysis, the subjective perceptions and observations were gathered and considered by the project team on the bases of the varying perceptions, in order to objectify them through a lengthy communications process. A common perspective of varying observations phenomenon was hereby developed, and a selection of themes was made which appeared, from an external perspective, to be relevant to this social context as well as the questions underpinning the project. This evaluation process resulted in a total of 14 hypotheses which provide a thematic insight into the field of residential care for people with dementia. These are presented in this book chapter as examples only. Therefore, in addition to the selected *theses (indicated by italics)*, the underlying situational descriptions as well as an evaluation of discussions[7] with residents (B), community service volunteers (E), relatives and nursing staff (P) was illustrated. The following are introduced as examples of the themes: a) the discrepancy between care and support, b) training requirements for supporting people with dementia and c) neighbourhoods as social spaces for people with dementia.

The Discrepancy between Care and Support

A series of situations in the residential area of the home led observers to the conclusion that due to time restrictions and institutional guidelines, such as the categorisation of task fields into so-called 'care stages', little or no support can be given or adapted to the individual person and their individual personality. The nursing staffs' tasks focus predominantly on straight forward care activities.

Situation (exemplarily, observations protocol 2): a B screams "Let me go, damn it. I haven't broken anything, have I. I'm not in prison. Ordered onto the toilet. What nonsense. How much longer do I have to sit here?' Another B comes out of her room without a Zimmer frame, therefore uses the handrail, care trolley and plate trolley instead. "I can't go any further!" P walks past. B is in a wheelchair by the window but can't see out, and due to her physical condition is unable to move herself in the wheelchair. When asked if she wants to attend the programme, she responds with a definitive no, but is taken there in her wheelchair, regardless.

---

7 | In relation to this, it should be noted that the discussions were not initiated by the observers, rather contact between the residents of the care home and the observers was actively sought by the residents.

There is general consensus that the daily routine in the care homes that were investigated follows a set pattern which determines the activities of both the care providers and those receiving the care. Getting up in the morning, washing and meals as well as the activities on offer provide a certain amount of structure to the care providers' work flow. A legally stipulated ratio determines the number of registered nursing staff to those in need of care, and the care provision may only be provided by this nursing staff, whilst housekeeping activities may be carried out by porters. Under normal circumstances, this process operates smoothly. However, when incidents occur, for example if a resident wets their bed, falls over or is in a depressed state of mind, and other demands are being made, great strains are placed on this process. The entire nursing staff identified time pressures and stress as shortcomings, as well as the feeling that they were able to tend to the physical needs of the elderly people whilst conversation or support activities were not recognised. From the point of view of those receiving care, this shortcoming was expressed predominantly by the relatives. The people with dementia had few opportunities to comment, having to comply as part of the process. Extraordinary nervousness or an apathetic demeanour was observed most probably as a reaction to these pressures. Thus a distinctive discrepancy was identified between physical care and emotional/mentally stimulating care.

This conflict is highlighted by an example of two quotes taken from conversations (community service volunteers group interview):

E1: "The residents are still spending too much time sitting around their rooms, alone. The bed-bound are provided for, but that's it." E2: "You can't put people in their wheelchairs at 8am, and not take them eat until midday."

These quotes indicate that despite a distinct sense of responsibility as well as a high level of competence among the nursing staff, the structural framework leaves little room for the provision of holistic care for people with dementia.

Training Requirements for supporting People with Dementia

Despite deepening research and medical understanding of the progressive stages of the disease dementia, this understanding was barely transferred into care in the residential unit investigated. This was established as a shortcoming within the framework of the observation, and provided a thesis which characterised many subsequent situations: *Professionalization within the framework of the dementia syndrome is shaped, to a certain degree, by training such as building a biography, or acquiring a deeper understanding of the dementia sufferer's life story. This additional learning is necessary in order to sufficiently understand on the one side the dementia sufferer's behaviours, and on the other, in order to adequately handle and react to these behaviours.* It was found that despite an overwhelming majority of residents suffering from dementia, only the registered nursing staff had taken the opportunity for further study in relation to the care of people with dementia (biography building). This nursing staff was, however, engaged in 1:1 day care support in the facility, and not present in either residence. Thus it emerged that due to a lack of training amongst the nursing staff regarding the syndrome dementia, specific situations in the everyday care were either unresolved, or unresolved in the long term.

Situation (exemplarily, observations protocol 2): B comes out of her room and drops her key off with a P. "I'm going home". She then walks in the direction of the lift and leaves the care home. In the residence there is great consternation, as it is not known that the resident is only going to her son's house, a couple of streets away, and is then coming back alone. The son is informed. A member of the nursing staff reports that in such situations the police are also informed.

Situations (exemplarily, observations protocol 1 and 2): There is a resident who continually screams:

B: "Mother, Mother, Mummy, Mummy. Who's saying that?" Rolls back and forth in the wheelchair. Appears agitated. "It's too much for me. Come here. Come here. Mummy."

B is collected by P. P begins to feed her straight away. B screams and so is taken immediately to another lounge.

B screams "Lunch. Meat. Meat. Please. I shouldn't say it. Please (ten times). Don't tell anyone. Mummy. Cheating. Coffee cup. It's a cheat. Uncovered."

P thereupon says definitively "Right, that's enough!" and takes B to her room

Situation (exemplarily, observations protocol 4). A resident sings constantly. B comes out of her room singing and puts the chairs in the eating area in order. She sings 'Little Hans', a nursery rhyme. Her singing gets louder. Other residents tell the observers, "She sings all day. Has done for years."

These two behaviours led in a large number of cases during the observations to a significant disturbance among the other residents, some of whom asked vociferously for quiet. In many cases, the thought and action processes of the dementia sufferers conflicted on the one hand with the facility's structural processes. On the other hand, the dementia sufferers' statements and phrasing were often met with incomprehension, and were not taken in a wider context. It appeared however, that the dementia sufferers' diverse styles of expression could translate into biographical knowledge, leading to less incomprehension on the part of the nursing staff as well as the residents. Thus an adequate biographical study could provide solutions for such challenging situations.

Neighbourhoods as Social Spaces for People with Dementia

Within the framework of the observation, the residents were also observed on walks, and whilst shopping. This was of great interest as, ideally, the integration of people with dementia in neighbourhoods fits parallel to increased mobility. Technological access should create scope for enablement in the long term, in order that the neighbourhoods may be used both more and in an individual way. On the basis of observations, the notions of mobility, and thus neighbourhoods as a social space to be used by people with dementia must be revised. The thesis developed in this context frames the following observation: *The notion of neighbourhoods, within the framework of the research question, is a very individual social space. It consists of no more than a radius of 150m around the care home. Furthermore, amongst people with dementia, the desire for mobility does not seem to incorporate a desire to be outside. These residents identify the home as the social space. Thus the theoretical expectations of being outside and mobility were vastly different from the reality in the care home investigated.*

Although many residents make use of the home's open door policy and it's good location next to a park with a stream, only one in 50 residents goes on walks alone, taking responsibility for themselves. The length of the walk depends on their condition on the day, and the weather. On a good day the resident takes the longer route of about 500m; on less good days, and when it's raining or cold, the shorter route of around 250m was selected. In addition there was only one other resident who left the home alone and independently during our observation. She made use of the shopping options which were between 60m (the first shop) and 160m (the last shop) away. She takes a direct route from the care home to the shop or shops, does her shopping, and returns immediately, using the direct route. Her leaving the home to go shopping, however, was based on compulsive shopping caused by neurosis, as the nursing staff informed us.

Situation (exemplarily, observations protocol 2): B sets off with a shopping basket, in the bakery asks other customers for 1 euro. Sales assistant comes out: "Leave my customers alone immediately – go away!" The sales assistant ejects B from the shop.

The notion of integrating people with dementia into the neighbourhoods more, and making use of attractions in the neighbourhood in order to mobilise people presented itself as complex within the unit investigated. This is due in large part to the fact that the residents are mentally and physically no longer in a position to decide upon and carry out excursions in the neighbourhood. In addition, the social integration of people with dementia into the neighbourhood as a social space is also very complicated and it would have to be prepared and promoted using targeted strategies and publicity work. Although this objective was one of the main concerns when the facility was established, the expectations, according to the former head of the facility, are largely unfulfilled. Targeted political and social efforts must be made to promote the integration of people with dementia into neighbourhoods. This is, however, currently widely lacking.

7. KEEP MOVING!

The three example theses form only a fragment of the on-going research process presented, however they do clearly reveal two aspects. Firstly, there are huge areas of conflict between the expectation that residential care-providing organisations have an institutional structure and the expectation that care-provision includes the nurturing of people with dementia in the care facility's daily routine. These areas of conflict are widely known and have been debated within the field for a long time. Due to economic considerations, obvious solutions in residential care such as, for example, increasing the ratio of support workers or the comprehensive promotion of professionalization strategies within the care sector seem unlikely.

Secondly, technologies and visions for the implementation of technologies in care are, in the project presented, not necessarily at the forefront in the minds of the nursing staff, the people with dementia or their relatives.

However, the needs-oriented approach to technology development, particularly in care, has been promoted for a considerable time now, and user involvement in the development of innovative technologies in care are increasingly required. Most importantly, against a backdrop of increased use of Responsible Research and Innovation (RRI)[8], responsible technology development should also be translated into practice.

The case presented, however, shows that with the user group encountered (people with dementia) this approach could and was only applied in a limited way. This was due to the fact that very elderly residents lived in the care home selected, up to the age of 96, and that the majority of these, with few exceptions, were in the later stages of dementia. An initial and provisional conclusion can be drawn from the approach, that the method of assessing need must be adjusted for each individual and the syndrome. How the method is to be adjusted to the group of people with dementia requires further research and discussion. Adequate technologies for care and support, as well as danger-free access to individual social spaces in neighbourhoods as well a social life for people with dementia as 'protected participants' can only be developed with an optimal needs analysis.

Nevertheless, based on observations, it can also be determined that the multiple areas of tension in residential care may not be resolved just by the use of technology alone. It would surely be helpful to think here of social innovations, which would frame the future demographic transition in terms of social and communal functions.

---

8 | "Responsible research and innovation is an approach that anticipates and assesses potential implications and societal expectations with regard to research and innovation, with the aim to foster the design of inclusive and sustainable research and innovation. Responsible Research and Innovation (RRI) implies that societal actors (researchers, citizens, policy makers, business, third sector organizations, etc.) work together during the whole research and innovation process in order to better align both the process and its outcomes with the values, needs and expectations of society. In practice, RRI is implemented as a package that includes multi-actor and public engagement in research and innovation, enabling easier access to scientific results, the take up of gender and ethics in the research and innovation content and process, and formal and informal science education." (European Union 2015)

Acknowledgements

The authors would like to thank staff and residents from the participating residential care facility who made this work possible through their involvement, as well as the care home management for allowing and supporting the research. We are also grateful for the support and contribution of Johannes Hirsch, Silvia Woll and Marcel Krüger. This work has been undertaken as part of the Movemenz project "Mobile, self-determined living for people with dementia in neighbourhoods" which is funded by the German Federal Ministry of Education and Research (BMBF).

References

Abt-Zegelin, A./Reuther, S. (2011): "Bewegungsförderung. Mobil im Pflegeheim." In: Die Schwester. Der Pfleger 50/04, pp. 322–325.

Bachman, D.L./Wolf, P.A./Linn, R.T./Knoefel, J.E./Cobb, J.L./Belanger, A.J./White, L.R./D'Agostino, R.B. (1993): "Incidence of dementia and probable Alzheimer's disease in a general population: The Framingham study." In: Neurology 43/3, p. 515–519.

Baldewijns, G./Debard, G./Mertens, M./Devriendt, E./Milisen, K./Tournoy, J./Croonenborghs, T./Vanrumste, B. (2013): "Semi-automated Video-based In-home Fall Risk Assessment." In: Encarnação, P./Azevedo, L./ Gelderblom G.J. (eds.): Assistive Technology: From Research to Practice, Amsterdam: IOS Press, pp. 59-64.

Blättel-Mink, B. (2006): Kompendium der Innovationsforschung, Wiesbaden: VS-Verlag für Sozialwissenschaften.

Bias, R.G./Mayhew, D.J. (2005): Cost-Justifying Usability – An Update for the Internet Age, San Francisco: Morgan Kaufmann.

Blinkert, B./Klie, T. (2004): Solidarität in Gefahr: Pflegebereitschaft und Pflegebedarfsentwicklung im demografischen und sozialen Wandel; die „Kasseler Studie", Hannover: Vincentz Network GmbH & Co KG

Blythe, M.A./Monk, A.F./Doughty, K. (2005): "Socially dependable design – The challenge of ageing populations for HCI." In: Interacting with Computers 17/6, pp. 672-689.

BMBF (Federal Ministry of Education and Research) (2013): „Die Verankerung des Themas „Verantwortungsvolle Forschung und Innovation" in der Europäischen Forschungspolitik und -förderung". Bericht. Ausschussdrucksache 17(18) 386, http://www.forschungswende.de/images/ PDF/ADrs%2017-386.pdf (last access: 27.05.2015)

BMFSFJ (Federal Ministry of Family Affairs, Senior Citizens, Women and Youth) (2007): „Möglichkeiten und Grenzen selbstständiger Lebensführung in stationären Einrichtungen (MuG IV). Demenz, Angehörige und Freiwillige, Versorgungssituation sowie Beispielen für „Good Practice"". München. http://www.bmfsfj.de/RedaktionBMFSFJ/Abteilung3/Pdf-Anlagen/abschlussbericht-mug4,property=pdf,bereich=bmfsfj,sprache=de,rwb=true.pdf (last access:13.04.2015).

Clement, A./van den Besselaar, P. (2004): Artful integraton. Interweaving media, materials and practices. Proceedings of the eighth conference on Participatory design, Toronto, Canada: ACM.

Cooper, R.G. (2002): Top oder Flop in der Produktentwicklung – Erfolgsstrategien: Von der Idee zum Launch, Weinheim: Wiley-VCH.

Daley, M.J./Spinks, W.L. (2000): "Exercise, mobility and ageing." In: Sports Medicine 29/1, pp. 1-12.

Deutsche Gesellschaft für Psychiatrie, Psychotherapie und Nervenheilkunde (DGPPN)/Deutsche Gesellschaft für Neurologie (DGN) (eds.) (2009): S3-Leitlinie „Demenzen", Bonn.

Encarnação, P./Azevedo, L./Gelderblom, G.J. (eds.) (2013): Assistive Technology: From Research to Practice, Amsterdam: IOS Press.

European Union (2015): Horizon 2020: The EU Framework Programme for Research and Innovation. (http://ec.europa.eu/programmes/horizon2020/en/h2020-section/responsible-research-innovation (last access: 13.04.2015)).

Federal Statistical Office (2015). Health. Data & Facts. Wiesbaden. https://www.destatis.de/DE/ZahlenFakten/GesellschaftStaat/Gesundheit/Pflege/Pflege.html (last access: 28.04.2015))

Fleming, J./Brayne, C. (2008): Inability to get up after falling, subsequent time on floor, and summoning help: prospective cohort study in people over 90. British Medical Journal 337: a2227.

German Alzheimer Society (2014): The most important. The epidemiology of dementia, Berlin. https://www.deutsche-alzheimer.de/uploads/media/PM_09042012_Neue_Zahlen.pdf (last access: 27.05.2015)

Gimskär, B./Hjalmarson, J. (2013): "A Test of a Walker Equipped with a Lifting Device." In: Encarnação, P./Azevedo, L./Gelderblom G.J. (eds.), Assistive Technology: From Research to Practice, Amsterdam: IOS Press, pp. 3-9.

Glende, S./Podtschaske, B./Friesdorf, W. (2009): "Senior User Integration – Ein ganzheitliches Konzept zur Kooperation von Herstellern und älteren Nutzern während der Produktentwicklung." In: VDE/VDI-IT/BMBF (eds.): Ambient Assisted Living Technologien, Anwendungen, Berlin, Offenbach: VDE Verlag, pp. 70-74.

Grunwald, A. (2011): Einführung in die Technikfolgenabschätzung, 2nd ed., Berlin.

Hjalmarson, J./Svensson, H./Glimskär, B. (2013): "Accessibility for Elderly using a Four-Wheeld Walker – an Interview and Observation Analysis." In: Encarnação, P./Azevedo, L./Gelderblom G.J. (eds.), Assistive Technology: From Research to Practice, Amsterdam: IOS Press, pp. 27-33.

Kangas, M./Vikman, I./Wiklander, J./Lindgren, P./Nyberg, L./Jämsä, T. (2009): "Sensitivity and specificity of fall detection in people aged 40 years and over." In: Gait Posture 29/4, pp. 571-574.

Krings, B.-J./Böhle, K./Decker, M./Nierling, L./Schneider, C. (2013): "Serviceroboter in Pflegearrangements." In: Decker, M./Fleischer, T./Schippl, J./Weinberger, N. (eds.): Zukünftige Themen der Innovations- und Technikanalyse. Lessons Learned und ausgewählte Ergebnisse. KIT Scientific Reports, Karlsruhe: KIT Scientific Publishing, pp. 63-121.

Lauriks, S./Reinersmann, A./van der Roest, H./Meiland, F.J.M./Davies, R.J./Moelaert, F./Mulvenna, M.D./Nugent, C.D./Dröes, R.M. (2007): "Review of ICT-based services for identified unmet needs in people with dementia." In: Ageing Research Reviews 6/1, pp. 223–246.

Lenker, J.A./Harris, F./Taugher, M./Smith, R.O. (2013): "Consumer perspectives on assistive technology outcomes." In: Disability and Rehabilitation: Assistive Technology 8/5, pp. 373-380.

Löfqvist, C./Nygren, C./Brandt, A./Iwarsson, S. (2009): "Very old Swedish women's experience of mobility devices in everyday occupation: a longitudinal case study." In: Scandinavian Journal of Occupational Therapy 16/3, pp. 181-192.

Löfqvist, C./Nygren, C./Iwarsson, S./Szeman, Z. (2005): "Assistive devices among very old persons in five European countries." In: Scandinavian Journal of Occupational Therapy 12/4, pp. 181-192.

Meiland, F./Dröes, R.-M./Sävenstedt, S./Bergvall-Kåreborn, B./Andersson, A.-L. (2010): "Identifying User Needs and the Participative Design Process." In: Mulvenna, M.D./Nugent, C.D. (eds.): Supporting People with Dementia Using Pervasive Health Technologies, Berlin: Springer, pp. 79-100.

Milisen, K./Detroch, K./Bellens, K./Braes, T./Dierickx, K./Smeulders, W./ Dejaeger, E./Bonnen, S./Pelemans, W. (2004): "Falls among community-dwelling elderly: a pilot study of prevalence, circumstances and consequences in Flanders." In: Tijdschrift Voor Gerontologie en Geriatrie, 35/1, p.15-20.

Nygard, L./Starkhammar, S. (2007): "The use of everyday technology by people with dementia living alone: Mapping out the difficulties." In: Aging and Mental Health 11/2, pp.144–155.

Orpwood, R./Gibbs, C./Adlam, T./Faulkner, R./Meegahawatte, D. (2005): "The design of smart homes for people with dementia: User interface aspects." In: Universal Access in the Information Society (2005) 4, pp.156–164.

Pape, T.L.B./Kim, J./Weiner, B. (2002): "The shaping of individual meanings assigned to assistive technology: A review of personal stakeholders." In: Disability and Rehabilitation 24 (1-3), pp. 5-20.

Phills, J. A./Deiglmeier, K./Miller, D.T. (2008): "Rediscovering social innovation." In: Stanford Social Innovation Review 6/4, pp. 34–43.

Przyborski, A./Wohlrab-Sahr, M. (2010): Qualitative Sozialforschung. Ein Arbeitsbuch, Oldenbourg Wissenschaftsverlag, München.

Reichwald, R./Ihl, C./Seifert, S. (2004): Kundenbeteiligung an unternehmerischen Innovationsvorhaben – Psychologische Determinanten der Innovationsentscheidung, München: TUM.

Roentgen, U.R./Gelderblom, G.J./De Witte, L.P. (2013): "Outcome Assessment of Electronic Mobility Aids for Persons Who Are Visually Impaired." In: Encarnação, P./Azevedo, L./Gelderblom G.J. (eds.), 2013 Assistive Technology: From Research to Practice, IOS Press, pp. 571-576.

Salminen, A.U./Brandt, A./Samuelsson, K./Toytari, O./Malmivaara, A. (2009): "Mobility devices to promote activity and participation: a systematic review." In: Journal of Rehabilitation Medicine, 41/9, pp. 697-706.

Sanders, E.B.N. (2002): "From User-Centered to Participatory Design Approaches." In: Designing and Social Science, New York: Taylor & Francis.

Schäufele, M./Köhler, L./Lode, S./Weyerer, S. (2008): "Menschen mit Demenz in stationären Pflegeeinrichtungen: aktuelle Lebens- und Versorgungssituation." In: Bundesministerium für Familie, Senioren, Frauen und Jugend. Möglichkeiten und Grenzen selbstständiger Lebensführung. Integrierter Abschussbericht, Berlin, pp. 169−232.

Schlick, C. M./Bruder, R./Luczak, H. (2010): Arbeitswissenschaft, Berlin: Springer.

Schrank, S./Zegelin, A./Mayer, H./Mayer, H. (2013): "Prävalenzerhebung zur Bettlägerigkeit und Ortsfixierung – eine Pilotstudie." In: Pflegewissenschaft 04/13, pp. 230-238.

Skymne, C./Dahlin-Ivanoff, S./Claesson, L./Eklund, K. (2012): "Getting used to assistive devices: ambivalent experiences by frail elderly persons." In: Scandinavian Journal of Occupational Therapy 19/2, pp. 194-203.

Taylor, R. (2008): Alzheimer und Ich. Leben mit Dr. Alzheimer im Kopf, Bern: Hans Huber.

Weinberger, N./Decker, M./Krings, B.-J. (2014): "Pflege von Menschen mit Demenz – Bedarfsorientierte Technikgestaltung." In: Schultz, T./Putze, F./Kruse, A. (eds.), Technische Unterstützung für Menschen mit Demenz, Karlsruhe: Kit Scientific Publishing, pp. 61-74.

Wesley, F./Antadze, N. (2010): "Making a Difference. Strategies for Scaling Social Innovation for Greater Impact." In: The Innovation Journal 15/2, pp. 2-19.

World Health Organization (2010): Global Recommendations on Physical Activity for Health. WHO Press, Geneva, Switzerland

Zegelin, A. (2010): Festgenagelt sein. Der Prozess des Bettlägerig-Werdens, Bern: Hans Huber.

Emotional Robotics in the Care of Older People: A Comparison of Research Findings of PARO- and PLEO-Interventions in Care Homes from Australia, Germany and the UK

BARBARA KLEIN, GLENDA COOK, WENDY MOYLE

1. INTRODUCTION

Demographic change is an issue affecting most societies in the world although the ratios of older to younger members of society do vary: Australia with 14.7% and the UK with 17.3% of the population 65 years and older are slightly younger than Germany with 20.9% (CIA 2014). Demographic changes for these countries indicate a major increase, especially in the age group 80 years and older. Whilst it should not be assumed that all older people have dementia, the association between dementia and advancing age must be acknowledged. A recent Alzheimer's Disease International report (Alzheimer's Disease International 2010) indicated that more than 35 million people worldwide have dementia. This number will continue to increase with the ageing of the population since, after the age of 65, the incidence of developing dementia doubles every additional five years (Alzheimer's Disease International 2010).

It has consistently been reported in the research literature that older people who live in care homes (also referred to in Australia as nursing homes or residential aged care) experience social and emotional isolation (McKee/Harrison/Lee 1999; Hubbard/Tester/Downs 2003): a factor that contributes to these outcomes is the quality and type of social interaction within a care home. This is influenced by the personal attributes of residents, which include sensory deficits, communication, mobility and cognitive abilities. The physical environment of a care home and its cultural attributes – such as the philosophy of care and interventions implemented by staff – can also facilitate or inhibit social interaction between residents, staff and visitors. In recognition of these issues, there has been an increasing body of research and associated interventions that aim to enhance opportunities for social interaction (Dunn et al. 2010).

"Traditional" care provided by family members in many countries is still the most common form of care provision, although due to socio-demographic changes, the numbers of available family members who can provide such care are expected to decline. In the last few years, a variety of "new care concepts" have been developed, such as sheltered housing, new living arrangements in flats/apartments, shared communities in combination with extramural care, and the utilization of new technologies. So far the dissemination of these forms of care is still new.

Emotional robots such as the robot seal PARO or the dino robot PLEO were developed in order to stimulate emotions and thus they have the potential to initiate social interaction between the person with dementia and the robot and / or the caregiver. This type of robot is also called a companion type robot (Broekens/Heerink/Rosendal 2009) or sociable robot (Kidd 2008). The authors of this paper choose to use the term "emotional robots", with the assumption that the robots appeal to and evoke emotional feelings regardless of the person with dementia's age and illness. Due to their highly imitative and life-like behaviour, such robots can raise ethical concerns about deceiving the person with cognitive impairment. However, these robots have advantages over living animals as they do not incur vet fees and the stress placed on staff of feeding and walking an animal. Moreover, hygiene is minimal and interaction can occur without the presence of a carer and without the fear of the animal becoming stressed or causing injury to residents.

Libin & Libin (2002) defined emotional robots as a research area focusing on the analysis of person-robot-communication "viewed as a complex interactive system, with the emphasis on psychological evaluation, diagnosis, prognosis and principles of non-pharmacological treatment." (907). Since then, a number of pilot projects have been carried out in order to analyse the effects of different artefacts of emotional robotics (reviews: Broekens/Heerink/Rosendal 2009; Bemelmans et al. 2012; Kolling et al. 2013). This contribution looks at three different approaches, all of which utilize emotional robots: PARO[1], the therapeutic seal developed by AIST in Japan which is

---

1 | PARO is an artificial intelligence emotional robot in the form of a baby harp seal, which was designed to interact with human beings to elicit an emotional

utilized in Australia, Germany and the UK, as well as PLEO[2], a Camarasaurus dinosaur which has been developed as a toy.

PARO and PLEO are both emotional robots with different purposes: whereas PARO is especially designed for "therapeutic" purposes, PLEO is designed as a toy and it is therefore less robust when compared to PARO.

This paper discusses a PARO group intervention and outcomes observed in an Australian[3] and UK[4] care home (see also Moyle et al. 2013a; Cook/Clarke/Cowie 2009) together with the findings of teaching research projects in Germany[5] using PLEO (see also Klein 2011, 2012). The Australian research was undertaken in 2011; the UK research in 2009, and the teaching research projects in Germany were undertaken from the summer term 2009 to the winter term 2010/11. These research projects aimed to explore whether emotional robots could contribute to quality of life of people with dementia living in nursing home care. The methods varied and this chapter explores the project outcomes in order to achieve a deeper understanding of necessities of further research of emotional robots.

---

attachment to the robot (Wada/Shibata 2007). PARO has multiple sensors and a set of behavior action sequences: sensors include touch sensors over the robot's body, an infrared sensor, stereoscopic vision and hearing. Actuators include eyelids, upper body motors, front paw and hind limb motors. These sensors "recognize" behavior and trigger emotional states, while they provide the opportunity for the person to communicate with the PARO and the PARO to return the communication (Wada/Shibata 2007).

2 | PLEO is equipped with a camera-based vision system, microphones, beat detection in order to dance, touch sensors over its body, foot sensors for surface detection, a tilt sensor, infrared mouth sensors for object detection placed in the mouth, infrared communication with other PLEOs, and infrared detection for external objects (http://en.wikipedia.org/wiki/Pleo: 01.02.2015). The PLEO manual states that PLEO is a new life form, as it starts life as a baby and can develop its behavior to an adult dinosaur.

3 | The Dementia Collaborative Research Centre- Consumers and Carers funded the Australian study.

4 | The UK study was funded by DH Care Services Improvement Partnership.

5 | Frankfurt University of Applied Sciences funded a PARO and two PLEOs. The teaching research projects took place in regular courses in the Bachelor of Social Work and were therefore unfunded.

2. DIFFERENT APPROACHES TO THE EFFECTS OF EMOTIONAL ROBOT THERAPY

2.1 The Australian Approach

Practice Development in Australia
The Centre for Health Practice Innovation (HPI) at Griffith University aims to find solutions to critical healthcare challenges and to undertake cutting edge research that results in better health, better community care and improved quality of life for patients and clients. The Centre runs a randomization service and the majority of the research undertaken is by means of controlled trials with the aim of informing evidence-based practice. The Laboratory for Assistive Technology and SociAl Robotics (LASAR) was established in HPI in 2013 and is a state of the art social robotics laboratory that enables HPI researchers to bring older people, people with dementia and carers into the lab to evaluate and develop new equipment and software. The laboratory has a one-way screen and sophisticated monitoring systems allowing participants to be observed and recorded during evaluation of robots, assistive technology and software. As well as a significant number of social robots and assistive technologies, the laboratory has a video coding laboratory and software that enables video coding to take place. The laboratory also offers a training ground for health students and postdoctoral researchers. One of the key research foci of the ageing and older people research team in HPI is improving quality of life for older people with dementia living in nursing homes through encouraging social engagement, and one area of research has been the use of emotional robots such as PARO.

The majority of nursing homes in Queensland are either non-profit or private institutions that can make decisions about whether to be involved in research or not. The researchers sought interest in being involved in research involving PARO from two large nursing home providers. One provider declined, as they viewed the robots as infantilizing older people, whereas the participating provider was interested in improving quality of life for people with dementia and viewed the PARO as offering this opportunity.

Aim
The aim of the pilot project PARO was to seek data on the effectiveness of PARO in engaging people with dementia. The researchers aimed to look at the feasibility of using PARO and, if successful, to use the data to seek further fund-

ing to undertake a large multicenter cluster-randomized controlled trial. The researchers were recently successful in receiving funding and commenced a large cluster randomized controlled trial (c-RCT) in 2014 (Moyle et al. 2013b).

Methods

The pilot study compared the effect of PARO (intervention) to participation in an interactive reading group (control) on emotions in people living with moderate to severe dementia in a nursing home setting. A randomized crossover design with PARO and reading control groups was employed. A reading control group was chosen, as this was a usual activity used within the care home to engage groups of people with dementia in a social activity. The reading group engaged the residents in similar activities used within the PARO intervention (as outlined below). Eighteen people with mid to late stage dementia were recruited for the study.

A trained facilitator undertook both the intervention and control activities for 45 minutes, three afternoons a week, for five weeks. Participants then crossed over into the opposite activity and the protocol was repeated following a three-week period of no activity (washout) (Moyle et al. 2013a). The intervention and control activity were undertaken in a small group of participants (n=9). The researchers drew on the descriptive PARO research of Cook (Cook,/Clarke/Cowie 2009) (see below) in designing the PARO and control intervention, designing the PARO intervention around the following concepts: discovery (examining PARO); engagement (encouraging participants to talk and touch PARO); social interaction (the facilitator encouraged questions about PARO to be discussed in the group); and touch (touching and describing the fur or PARO's eyes). One PARO was introduced in week 1 to 3 and two PARO were introduced to the group in week 4 to 5. The reading group also followed the same processes but concentrated discovery, engagement, social interaction and touch on the stories being read by the facilitator. The facilitator was an arts graduate with experience in conducting activity therapy with people with dementia. The lead researcher trained the facilitator, while the lead researcher and one other team member oversaw the conduct of the intervention.

Outcome measures were undertaken at three time points: baseline (pre-intervention), mid-point (after the first 5-week intervention arm) and post-intervention (after the second 5-week intervention arm). The primary outcome measure was quality of life using the Quality of life in Alzheimer's Disease Scale (QOL-AD, a modified version for use in a nursing home population)

(Edelman et al. 2005). Mood states were measured with following secondary outcome measures: Geriatric Depression Scale (Yesavage 1988); Observed Emotion Rating Scale (OERS); (Lawton/Van Haitsma/Klapper 1999) Apathy Evaluation Scale (Marin/Biedrzycki/Firinciogullari 1991); and Algase Wandering Scale-Nursing Home version (Algase et al. 2001) The researchers also video recorded one session each week and these were analyzed using Noldus software for engagement and emotional response. The research was funded by the Dementia Collaborative Research Centre-Carers and Consumers.

Findings
The findings have been previously reported (Moyle et al. 2013a) and therefore this paper will provide a brief summary of the findings. The overall findings were positively in favor of PARO when compared to the reading group. PARO was found to have positive, medium- to large-effect sizes on the QOL-AD (0.6 to 1.3) and OERS pleasure subscale (0.7) in the PARO group: these scores were higher than in the reading group. The Noldus video analysis also suggested that participants in the PARO group displayed less anxiety than those in the reading group. They also displayed longer periods of positive engaging behaviors during the PARO sessions such as looking directly at PARO, smiling, laughing, touching and talking to PARO.

All sessions were conducted in small groups (n=9). As indicated above, in the first three weeks the ratio of the PARO was 1:9 and in the remaining two weeks it was 2:9. The large group size reduced the amount of individual time participants could have with PARO and this negatively influenced participants' wandering behaviors. For example, when two or three group members were engaged with PARO and the facilitator was facilitating their discussion, there were times when some of the remaining individuals lost interest in the activity. When one resident got up and wandered aimlessly around the room, or at times out of the room, this distracted the group from the PARO activity. The researchers perceived that PARO may be more therapeutic in a one on one situation rather than a large group situation. The current large cluster randomized controlled trial (c-RCT) uses one PARO with one resident.

The pilot data had some surprising findings, such as the fact that individuals classified by staff as being non-communicative began speaking to PARO, asking questions and making statements about it. Most of these statements were part of their engagement with PARO. They would address PARO in ways such as: "You are beautiful. Your eyes are lovely". Although the findings were generally positive, the researchers advocate for the need for a larger study to help determine whether PARO is, indeed, a short term, low risk, non-pharmacological intervention that produces tangible positive psychological outcomes for people with dementia. Further research must also consider a comparative cost analysis to determine if PARO is as cost-effective as a pharmacological intervention or an alternative activity such as music therapy, social activity with a volunteer, or cheaper alternative robotic pets/toys. The current cluster randomized controlled trial is undertaking a cost analysis.

2.2 Teaching Research Projects in Germany

Since 2009 the Faculty of Social Work and Health of the Frankfurt University of Applied Sciences has used emotional robots such as the therapeutic seal PARO, and since 2010 two toy dinosaurs PLEO in teaching research projects in the Bachelor Degree program in Social Work (Klein 2011, 2012). Students are taught the theoretical concepts of socio-pedagogic approaches in nursing care homes and they have to develop a concept for assisted activities with new technologies.

Aims
Objectives linked with the teaching research projects are that students get into contact with their future clients, transfer theoretical knowledge into practice, develop their observational skills and explore the potential of new technologies for daily activities.

Methods
Artefacts such as the therapeutic seal PARO or the toy PLEO are implemented in teaching research projects in a module on "client-orientation and well-being in service provision of elder care". In the course module, students deal with social work in elder care and learn a variety of methods and tools for daily activities. Based on that knowledge, they have to develop an activity concept for a minimum of three sessions and implement it in a nursing care home: such sessions can be based on robot-assisted activity. Their observational skills on the effects of the intervention are developed – they have to videotape the sessions, analyse their videos, write a report on their observations and experiences, and reflect on their effects on the wellbeing of the residents.

Teams of three to five students have to carry out the project within four weeks. After having obtained informed consent of residents (or their legal custodians) and the management, they facilitate at least three sessions with residents in nursing care homes with the selected technology. Afterwards, they report the results and have to do a project presentation.

Findings

The course takes place twice a year. In the period between the summer term of 2009 and the winter term of 2010/11, there were a total of eleven robotic interventions in different nursing care homes: seven groups used PARO and four groups chose PLEO. Due to quality issues, only six project reports on PARO are taken into account in this chapter.

During this time period, a total of 62 residents had contact with emotional robots; 88.7% were female, which corresponds to the average sex distribution in nursing care homes. 38 of the residents had activities with PARO, 86.8% of which were female; and 24 had activities with PLEO, 91.6% of which were female.

Students undertook both group and individual interventions. Group size varied up to ten residents, as findings suggest that a group size of up to four residents can be managed more easily. Some of the students were rather skeptical towards the use of robots for interventions, but their experiences resulted in a change of their attitudes; thereafter, the students often saw potential for robot activities. Three persons out of 38 with PARO interventions did not like the seal; one person left the group intervention. In the individual interventions, two residents refused PARO by showing their dislike, either by shaking their heads or saying no. Students were instructed to respect wishes of residents not to participate in robot interventions and not to question this decision even if there is a written agreement in advance.

The findings of the reports indicate that emotional robots stimulate (positive) emotions and social interaction, most times in a positive way. The analysis of the German project reports revealed the reactions to the emotional robots described below and the opportunities the robots offered to older residents. The categorization was obtained by listing the activities students mentioned and then categories were derived, which represent qualitative different levels of social interaction:

- *Mimic expressions and gestures.* Residents expressed and gestured at the robot: They looked at the robot, but also to other persons in the room and communicated via grinning, smirking, smiling, laughing. These observations often had explanations such as "resident does not usually smile"; "resident does not usually show such positive emotions and happiness",

indicating that the positive reaction resulted from the introduction of the robot (Klein 2011).
- *Touching the robot.* This included stroking, cuddling and hugging. In one of the project reports, these interactions with intense skin contact are interpreted as a new basal stimulation approach (Bienstein/Fröhlich 2010) to people suffering from dementia, which might contribute to either reducing aggressiveness or to stimulate a positive mood.
- *Verbalization / talking to the robot.* Holding and touching the robot were accompanied by talking to the robot. The way residents talked to the emotional robots was seen as being similar to the way adults talk to babies and toddlers – with higher intonation and of confirmative or asking character.
- *Stimulation of social interaction.* In a similar way to the situation in England (see below), social interaction between residents is not taken for granted. Even in activity sessions, communication structures can be restricted only between residents and facilitator. However, the reports revealed examples where the emotional robots encouraged discussions between residents and, as happened with the British experiences, talks were on pets or memories of past times e.g. such as former vacations. One of the project reports mentions that two women with dementia started to talk about their health status and how horrible it is not to "recognize their own folks" or "remember the name of their husband".
- *To descend into their own world.* Two project reports observe that a resident became withdrawn in his or her own world and ignored the students and the robot, suggesting that people with severe stage dementia may not be able to display an emotional response towards the robot.
- *Caring behavior towards the robot.* In the individual robot interventions, the residents displayed caring and nurturing behavior towards the robot, such as getting a blanket to keep the robot warm or feeding PLEO with its (plastic) leaf.
- *Recreation.* PLEO displayed a range of activities beyond that of PARO. For example, PLEO is able to take little steps and its communication abilities are more developed. As a result, the residents enjoyed joining in singing with PLEO and they indicated that they enjoyed its robotic voice very much.
- *Dislike of robots.* Students were advised that if a person did not want to interact with the emotional robot, this had to be respected. The project reports mentioned that one person left the room or shook their head and said 'no' when asked if they would like to interact with the robot.

2.3 Practice Development in England

As part of a practice development initiative Northumbria University developed a framework known as INTERACT[6] to enhance social interaction with residents in nursing and residential care homes. The framework "raise(s) awareness of innovative ways of promoting social engagement" (Cook/Clarke/Cowie 2009: 5) The INTERACT Framework has been developed for health and social care staff and offers guidance and strategies to promote social interaction in the resident population. Using this framework in an exploratory study, PARO was introduced to five residents. This activity specifically sought to enhance social interaction between older residents with dementia through a novel intervention that involved facilitated group discussion with the emotional robot PARO.

Aims
The aims of the study were to implement facilitated PARO group discussions with residents with dementia and to observe the effect of PARO with respect to conversation and social interaction.

Methods
This was an ethnographic study of facilitated group discussions with PARO in a care home in North-East England. The care home is a modern and purpose-built centre that comprises four units, each with 20 bedrooms. Each unit has a dedicated team of staff who provides different forms of care: the PARO group discussions took place in an Elderly Mentally Infirm (EMI) unit in the centre. In addition to bedrooms, this unit had a dining area, small and large communal lounges, and bathroom facilities. The philosophy of care was person-centred, giving priority to addressing individual needs and providing a stimulating activity programme, which included music (such as playing instruments and listening), art (making cards, drawing) and gardening.

The PARO group discussions were held in an afternoon for one and a half hours in a small lounge in the EMI unit; the door was kept open during

---

6 | INTERACT stands for I "Individualise type and quality of social interaction; N "Notice the quiet, withdrawn resident"; T "Time to talk"; E "Environmental conditions"; R "Recognise and support relationships"; A "Assess individual problems and Action plan"; C "Create the Care Home Community"; and T "Use Technologies to support interaction".

the session. The participants were five people with dementia (3 females and 2 males between 75 and 88 years of age) who had been residents in the EMI unit of the home between 2 and 10 months. Four participants had good communication skills and one had not spoken for some time. Two participants were wheelchair users.

The researcher facilitated the sessions and was supported by a care assistant who had known the participants for at least one year. The sessions were held for a period of five weeks.

- Session 1: orientation. The PARO was placed out of sight while residents entered or were assisted to the room and were seated around a table. At this point, the facilitator explained that they had brought something about which she would like their opinion. After this introduction, the PARO was brought out, placed on the table, and turned on. The residents were told 'I have brought something for you to see today. This is PARO. It was given to us by someone from Japan. I am curious about what you think of PARO.' After some introduction, PARO is held by each member of the group. As each participant held PARO, the facilitator asked them questions such as: 'What do you think of PARO? What do you want to know about PARO? What do you like or dislike about PARO?' The session ended when the discussion ceased and PARO was turned off. Participants were asked if they would like to take part in a discussion with PARO next week.
- Session 2: PARO was turned on when the participants were seated around the table. They were asked if they could recall the PARO discussions from the previous week and were invited to interact with PARO in any way that they wanted to. The facilitator led discussions about what name should be given to PARO. They were also invited to discuss the same questions as the previous week: 'What do you think of PARO? What do you want to know about PARO? What do you like or dislike about PARO?'
- Sessions 3-5: Following initial interaction with PARO and exploration of any issues that arose spontaneously, the facilitator introduced the following topics: 'Have you had a pet in the past? What type of pet? How long did you have the pet and what did you do with it? What memories do you have of the pet? What were the most memorable moments with your pet?' At the end of the fifth discussion, the participants were asked about their views of participating in the group discussions.

Data collection involved observing interaction between the participants and PARO, among each other, and between the group participants and others who were not part of the group (residents who came into the room and staff). Following each session, notes were made of the observation, which were validated by the supporting carer. When the sequence of discussions was completed, interviews were held with the carer who supported the facilitator and with other staff who had observed the PARO intervention when they walked into the room during sessions. A verbatim transcription was made of the interviews and thematic analysis was completed across both observation and interview data sets.

Findings

Prior to entering the small lounge, the participants were gathered in another nearby lounge. Little social interaction was observed between the nine residents in this room. When approached by the staff and asked if they wanted to join the PARO group, there was no hesitation and they quickly settled around the table. PARO was placed on the table and switched on. Attention was focused on PARO; some participants smiled and spontaneously commented about PARO. These were short exchanges such as 'Look at what it is doing;' 'Oh, it is so lovely;' and when PARO made sounds, they asked 'Is it ok?' 'What does it need?' When one participant put out their hand to stroke PARO, they were invited to hold it. As they stroked and held the robot, they kept eye contact with PARO and moved their head following its movements. Verbal interaction involved the participant making soothing comments to PARO – 'There, there;' 'Look at you, oh you like that' in response to PARO's squeaks. Other participants made strong eye contact clearly observing the interaction between human and robot.

After five minutes, other members in the group were invited to hold PARO, giving each person the opportunity to have close contact with the robot. The participant with advanced dementia stroked and cuddled PARO, and swayed back and forth as if she was rocking the robot. This behavior contrasted to her previous state where she appeared to doze, following her initial interest in PARO. Other participants throughout the whole session maintained interest in PARO, which was evidenced by their comments and questions: they wanted to know how it worked, what it needed, how much it cost. Two participants referred to PARO as a real animal, a dog, indicating that it might need to go to the toilet and that it should have a rest, suggesting they were familiar with this type of animal. This real/machine distinction was implicit in their questions rather than being a point of discussion. These types of interaction were witnessed throughout all of the PARO sessions.

In subsequent sessions, the participants were observed advising others in the group about how to interact with and care for PARO. They commented 'He needs to be stroked in this way;' 'He is upset, talk to him more.' They endowed PARO with a masculine gender and, when asked if they wanted to give PARO a name, they agreed that he should be called 'Jimmy'.

In addition to the five residents who agreed to take part in the group sessions, other residents showed interest in what was taking place and they entered the room, joined the group and participated in discussions that related to PARO. Their comments and non-verbal interactions were similar to the participants; they wanted to hold PARO and engaged in one-sided conversation with the robot.

In one session, the woman with advanced dementia was given PARO. She sat back in her chair and constantly patted the robot and smoothed its fur. When another resident spontaneously joined the group, he was given PARO. This woman opened her eyes and watched him sitting quietly talking to the robot saying 'There, there puppy', 'Quiet now puppy'. He was very gentle with the robot and constantly patted it. His dialogue continued with positive comments to the female resident saying 'You have a lovely puppy.' In response, the woman appeared animated and she did engage in three brief exchanges with the male resident, stating 'Yes he is lovely.' She maintained eye contact whilst talking to him and moved her body forward in a positive gesture. He did likewise and smiled in response to her comments. This appeared to be a lucid moment, since these individuals were positively interacting with each other. This brief interaction was followed by the woman sitting back in her chair and closing her eyes appearing not to engage with others in her surroundings. The man continued to make positive comments about PARO and then spontaneously stated that PARO ought to be returned to the woman. When he passed PARO back to her, she opened her eyes again, maintained contact with him, and then started to stroke PARO in a slow consistent way from its head to its tail. There was one other resident in the group at this stage and he observed the episode. When the woman was holding PARO again, he also commented that she had a good pet and advised her to enjoy this because all pets were not so good.

In contrast, one of the participants appeared to be upset by the presence of PARO during the third session. She did not want to hold PARO and mentioned that they were all in danger. When asked if she wanted to leave, she responded positively. As she left the room, she appeared less anxious. She was invited to participate in the following session: however, her non-verbal behavior did not indicate agreement and therefore did not return to the group.

The facilitator introduced different topics to the discussion following the initial orientation session. Group members spoke of their pets, often dogs and cats. They told stories about interacting with their pets and this led onto other discussions about what they liked/disliked about interacting with animals. They spoke of places that they had both visited and which involved their pets. This prompted further discussion between the participants about what they did in those places. For example, one man spoke of times in his youth when he walked greyhounds and had gone to the racing stadium. In another situation, the two men discussed the route where one of them had walked the greyhounds: this was past the coal mine that no longer existed. They had both worked down the pit and they discussed their work: both commented on the caged birds that they took down the pit to detect hazards.

Care staff spontaneously took time out of their activities and observed the group and this led to impromptu discussions initiated by them with the residents. They were very keen on finding out what the participants thought of PARO. Two members of the staff observed the interaction between the residents that was described above and, after the session, they stated that they longed for that brief exchange to continue. They indicated that it had been a while since the female with advanced dementia had reacted in this way, that she had seemed relaxed and had enjoyed the session. They indicated that introducing PARO into the care environment promoted social interaction between residents, and between residents and staff. It was a trigger to start conversations and interactions that did not otherwise take place.

3. DISCUSSION: INTERACTION THROUGH ROBOT THERAPY?

The projects in Australia, Germany and the UK are not readily comparable, as each used different methodologies and methods, although they all seem to show indications that participants readily interacted with the robot by demonstrating emotional feelings and positive social interaction. Examples of the outcomes are presented and analysed with respect to the indicators of interaction as an outcome of the robotic intervention and the impact of emotional robots on the enhancement of social interaction in care.

The facilitated PARO group interventions in Australia provided both entertainment and stimulated engagement in a majority of participants. However, individual responses to PARO were not consistent. In some sessions, there were individuals who were more engaged than others and individual response could be positive one day and negative in the following session. Some of this can be explained by the fact that the researchers were unable to control for variables such as prescribed medication that may have influenced mood and response, or other influencing factors such as staff and family influences. There is a need for a larger project that tries to control some of these potential influences. The current cluster randomized controlled trial in Australia is the largest social robotic trial worldwide to explore the effect of social robots on people with dementia: with a sample size of 380 participants, the trial will identify the effect of usual care with the therapeutic robot PARO and with a look-alike plush toy (without the artificial intelligence aspects of the PARO).

The following table provides an overview on the findings:

Table 1: Overview of the Projects' Results

Overview	Australia	Germany	UK
Year when projects were undertaken	2011	2009-2011	2009
n	18	62	5
Characteristics of the "research" use of PARO	Randomized cross over design with Paro and reading control groups	Teaching research projects in order to qualify students in observational skills and issues of social work	Ethnographic study of facilitated interaction with PARO
Information collection / Research methods	Outcome measures at baseline (pre-intervention), midpoint (after 5-week-intervention-arm), and post-intervention (after second 5-intervention arm)	Observing interactions, usually supported by videographing	Observing interaction
Instruments	QOL-AD, modified version	Additional: sociogram of the interventions, smiley-scale for residents: How did you like...?	Notes after each session, which were validated by the supporting carer
	Geriatric Depression Scale; Observed Emotion Rating Scale (OERS); Apathy Evaluation Scale; Algase Wandering Scale-Nursing Home version	Questionnaire or interviews with (nursing)care staff or social worker	Interviews with the carer who supported the facilitator and other staff who observed the sessions
Analysis	Video recording of one session each week, analysed using Noldus Software for engagement and emotional response	Write up in project reports	Verbatim transcription, thematic analysis across observation and interview data set
Findings			
Residents	QOL-AD: positive, medium to large effect sizes; 0.6-1.3	Students observed mainly positive reactions on PARO and PLEO	Residents were mainly positive
	OERS pleasure subscale: Cohen's D (standardized difference in means) = 0.7 in favour for the PARO group compared to the reading group	Reactions were activities such as touching the robot; mimic expressions and gestures	Touching, stroking, smiling, following the movements; eye contact, rocking the robot
	Video analysis: participants in PARO groups less anxiety; longer periods of positive engaging behaviours	Verbalisation and talking with the robot; stimulation of social interaction, reflection on the effects of the illness;	Verbal interaction: soothing, talking about Paro,
	e.g. looking directly at PARO, smiling, laughing, touching and talking to PARO	descending in their own world; caring behaviour, recreation with PLEO	Brief exchanges between residents, but also staff
Staff	Staff observed that non-communicative individuals started to interact	In some cases students reported that a resident started to interact and talk who had not done so for some time according to staff	Staff observed that PARO promoted social interaction between residents and between residents and staff. /Trigger to start conversations and interactions that did otherwise not take place
	Indications that group size of 9 should be minimised in order to enable all participants more time with PARO	Group sizes varied; 4 residents seem to be manageable	
Rejection of Paro	-	Dislike of PARO in 2 cases, residents did not continue their participation	One person got upset in the third session / removal from the group
Human / Machine distinction	-	Real / machine distinction was observed	Real / machine distinction was observed

In these three studies, the focus was to find out whether this new robotic tool had any effect on the participating residents in the nursing care homes. The studies were undertaken between 2009 and 2011. The methods are not comparable, as no common assessment instrument was used throughout the studies: it is also not possible to compare the demographic variables of residents and staff. However, all three studies reported that the participating residents reacted mainly positively towards the PARO intervention.

In all three studies a facilitator moderated the interaction with PARO. In Australia, they were trained employees of the nursing care home; in the UK, the researcher took up the role as facilitator; in Germany, students facilitated the project in order to gain experience. The role of the facilitator seems to be crucial for the design of the interventions such as handing PARO over to the resident, giving a fair share of time with PARO for each resident, initiating topics to talk about, etc. These experiences can be brought into practice development (Klein/Gaedt/Cook 2013). They also open up a variety of issues still under researched such as how long, how often, and in what intensity interventions should be designed in order to contribute to wellbeing.

4. Conclusions and Recommendations for Future Research

All three projects demonstrate a number of positive findings for using emotional robots. However, the projects also raise more questions than they answer.

In particular, the question is raised whether it is the robotic characteristics that help to engage interaction or whether it is the novel appearance of the robots. Furthermore, the question of whether robots can produce the same or more enhanced engagement than living animals needs to be addressed, as well as the question of whether a person or a stuffed (non-robotic) animal can produce similar outcomes. PARO and PLEO are more interactive than a stuffed toy. If the perception and acceptance of those emotional robots is comparable with living animals, they might be an additional choice or even an alternative for some nursing care homes. In all participating countries, social activities comprise a variety of choice (e.g. reading groups, cooking, sports, music, etc.) for residents, although residents may not be able to or want to participate in such activities. Emotional response robots can offer a new activity to individuals living in nursing care home and robot therapy can thus extend the range of interventions that can be used in a care setting. There is little doubt, however, that introducing robots into nursing homes has the benefit of a new and stimulating opportunity that may help residents during those inevitable times when they are left alone and with limited comforts around them. There is little question that the novelty of the robots can induce joy and pleasure in residents: opportunities for enjoyment can potentially increase quality of life – even if it is only for a short time.

However, there was also a small number of residents in the UK and Germany who rejected PARO. In the UK, one person got upset after the third of five sessions and left the group. In Germany, one person did not want to get involved in the PARO interactions. Neither in the UK nor in Germany are more details known on the causes why people rejected involvement with PARO. Future research should collect data on the causes of resident rejection of PARO.

Caring for people with advanced dementia can be a challenging task for staff and relatives: opportunities to engage people with dementia in social interaction and to do something that is pleasurable may thus also have a positive impact on staff and family. Therefore, further research must consider the impact of robots not only on the person with dementia, but also on those around them.

There is a need for studies involving larger numbers of participants and where conditions are controlled so that we can identify the effect of such robots. Gender and cultural background should also be considered in order to get more insight on their role in acceptance.

More work has to be done to develop conceptual frameworks and models on the causes why emotional robots impact on people with dementia. Ethical issues also have to be analyzed in more depth. Discussions such as "robot replaces human beings and real emotions" around the robot seal misconceive the actual capability and "skills" of this particular robot as practice experience in German care homes actually shows that the robot seal is used together with professional staff. Additionally, European sales and distribution go along with staff training in robot assisted activities or therapy. Such issues cannot be dismissed, as they are relevant for further emotional robotic development and acceptance on a wider scale.

REFERENCES

Algase, D.L/Beattie, E.R.A./Bogue, E.-L./Yao, L. (2001): "The Algase Wandering Scale: initial psychometrics of a new caregiver reporting tool." In: American Journal of Alzheimers' Disease and Other Dementias 16/5, pp. 141-152.

Alzheimer's Disease International: World Alzheimer Report (2010): The Global Economic Impact of Dementia. Alzheimer's Disease International (ADI). (last access: 21.09.2010).

Bemelmans, R./Gelderblom, G.J./Jonker, P./de Witte, L. (2012): "Socially assistive robots in elder care: A systematic review into effects and effectiveness." In: Journal of the American Medical Directors Association 13/2, pp. 114-120.

Bienstein, C./Fröhlich, A. (2010): Basale Stimulation in der Pflege, Bern: Huber.

Broekens, J./Heerink, M./Rosendal, H. (2009): "Assistive social robots in elderly care: a review." In: Gerontechnology 8/2, pp. 94-103.

CIA (2014): The World Factbook, https://www.cia.gov/library/publications/the-world-factbook/geos/gm.html (last access: 01.02.2015).

Cook, G./Clarke, C./Cowie, B. (2009): Maintaining and developing social interaction in care homes: a guide for care home, health and social care staff. Newcastle: Northumbria University.

Dunn, J./Balfour, M./Moyle, W./Cooke, M./Martin, K./Crystal, C./Yen, A. (2013): "Playfully engaging people living with dementia: searching for Yum Cha moments. American Psychological Association." In: International Journal of Play 3/2, pp. 174-186. Edelman, P./Fulton, B.R./Kuhn, D./Chang, C.-H. (2005): "A comparison of three methods of measuring dementia-specific quality of life: Perspectives of residents, staff, and observers." In: The Gerontologist 45/1, pp. 27-26.

Hubbard, G./Tester, S./Downs, M.G. (2003): "Meaningful social interactions between older people in institutional care settings." In: Ageing and Society 23, pp. 99-114.

Kidd, C.D. (2008): Designing for Long-Term Human-Robot Interaction and Application to Weight Loss. PhD. Boston: Massachusetts Institute of Technology.

Klein, B./Gaedt, L./Cook, G. (2013): "Emotional Robots. Principles and Experiences with Paro in Denmark, Germany, and the UK." In: GeroPsych 26/2, pp. 89-99.

Klein, B. (2011): "Anwendungsfelder der emotionalen Robotik. Erste Ergebnisse aus Lehrforschungsprojekten an der Fachhochschule Frankfurt am Main." In: JDZB (ed.), Mensch-Roboter-Interaktion aus interkultureller Perspektive. Japan und Deutschland im Vergleich, Berlin: Veröffentlichungen des Japanisch-Deutschen Zentrums Berlin, pp. 147-162.

Klein, B. (2012): "Robot-Therapy in Germany." In: The Society of Instrument and Control Engineers (SICE) 51, pp. 649-653.

Kolling, T./Haberstroh, J./Kaspar, R./Pantel, J./Oswald, F./Knopf, M. (2013): "Evidence and Deployment-Based research into Care for the Elderly Using Emotional Robots." In: GeroPsych 26/2, pp. 83-88.

Lawton, M.P./Van Haitsma, K./Klapper, J.A.(1999): Observed Emotion Rating Scale, www.abramsoncenter.org/PRI (last access: 01.05.13).

Libin Elena V./Libin, Alexander V. (2002): "Robotherapy: Definition, Assessment, and Case Study." Proceedings of the 8th International Conference on Virtual Systems and Multimedia, Creative Digital Culture, VSMM Society, Seoul, pp. 906-915.

Marin, R.S/Biedrzycki, R.C/Firinciogullari, S. (1991): "Reliability and validity of the Apathy Evaluation Scale." In: Psychiatry Research 38/2, pp. 143-62.

McKee, K./Harrison, G./Lee, K. (1999): "Activity, friendships and wellbeing in residential settings for older people." In: Aging and Mental Health 3/2, pp. 143–52.

Moyle, W./Cooke, M./Beattie, E./Jones, C./Klein, B./Cook, G./Gray, C. (2013a): "Exploring the Effect of Companion Robots on Emotional Expression in Older Adults with Dementia." In: Journal of Gerontological Nursing 39/5, pp. 46-53.

Moyle, W/Beattie, E/Draper, B/Shum, D/Thalib, L. (2013b): Effect of an interactive therapeutic robotic animal on engagement, mood states, agitation and antipsychotic drug use, NHMRC Project Grant APP1065320, 2014-2017.

Wada, K/Shibata, T (2007): Living with Seal Robots – Its Sociopsychological and Physiological Influences on the Elderly at a Care House. In: IEEE Transactions on Robotics. 23/5, pp. 972-980.

Yesavage J.A (1988): "Geriatric Depression Scale." In: Psychopharmacology Bulletin 24/4, pp. 709-711.

POLICY MAKING AND DISCOURSES OF AGEING

Navigating the European Landscape of Ageing and ICT: Policy, Governance, and the Role of Ethics

EUGENIO MANTOVANI AND BRUNO TURNHEIM

1. INTRODUCTION

Ageing and the use of Information and Communication Technologies (ICT) are two defining trends of our times. On the one hand, an ageing society is regarded in Europe as an indisputable part of social dynamics: by 2050, estimates predict that the elderly will account for 16 percent of the global population. Across EU Member States, the population aged 65 and over will continuously increase from currently 86 million to 141 million by 2050. According to Eurostat, the rise in the share of older persons will, in turn, lead to an increased burden on those of working age to provide for the social expenditure required by the ageing population for a range of related services (Eurostat 2010). On the other hand, ICT developments are pushing technological boundaries, redefining our social environments and the way we live: intelligent machines, monitoring systems and implants are increasingly likely to take on, at least in part, the process of care, support and even companionship.

In the European Union, Ageing and Technology, ICT and Ageing, or ICT for older persons[1] is seen as a growing market sector in its own right; it is also addressed as part of the larger portfolio of policies associated with ageing policy such as social security, employment, health care policy, etc. Within this broader context, ICT developments are portrayed as a hopeful promise with the potential to cut age-related health costs, improve caring services, and stimulate the social and economic contribution of the elderly to society.

---

1 | In this paper, these expressions are used interchangeably. For a discussion on terminology see VALUE AGEING (2013).

Technological innovations for ageing are reminiscent not just of technical choices; they are also mediated by supporting beliefs and values. Initiatives promoting Ambient Assisted Living, or ICT for living independently shake value frameworks and what we consider a good old age, good care conditions, personal autonomy, independence, and human dignity. In 2007, when the Action Plan 'Ageing Well in the Information Society' was adopted, the EU Commission acknowledged this, warning that "when ethical concerns are not addressed properly, they lead to a rejection [or low uptake] of technology solutions" (European Commission 2007a: 46, see also 2007b). Since then, the EU has made use of its control over research funds to stimulate substantial research about the ethical implications of technologies and ageing.

This chapter draws from the findings of an EU funded project in this area of research, the IAPP Marie S. Curie Action VALUE AGEING (VA). VA's main aim was promoting interdisciplinary and inter-sectorial research on the question of the "incorporation of fundamental values in ICT for Ageing" (VALUE AGEING, 2010-2014).[2] This contribution does not discuss values or how specific ICTs impact on the freedom, autonomy, or dignity of older persons.[3] This chapter purports to shed light on whether and how values are taken into consideration in the EU policy making with regard to ageing and technology (A&T).

In a first section, we offer a brief history of the EU information society policy with regard to older persons. In writing this review, we have been interested in the emergence of A&T as part of the evolving relationships between technology developments, demographic change, and societal dynamics. We focus on the content of those policies and the successive framings that they mobilise (ageing well, active ageing, healthy ageing, etc.) as affecting research and technology developments.

The second part looks at EU policy on A&T from a different perspective: the perspective of governance, narrowly defined as the actors, configurations and procedures available at EU level to expose and address ethical questions

---

2 | *VALUE AGEING: Incorporating European Fundamental Values into ICT for Ageing: A Vital Political, Ethical, Technological, and Industrial Challenge* is an European Commission FP7-funded Marie Curie Industry-Academia Partnerships and Pathways Action (Grant agreement no.: 251686; 2010 – 2014). The project aimed at fostering co-operation between non-commercial and commercial entities on a joint research project about the incorporation of fundamental values in Information-Communication Technologies (ICT) for ageing. www.value-ageing.eu

3 | Interested readers will find projects deliverables at www.value-ageing.eu.

with respect to ICT and ageing. This section draws on a constructive framework for the consideration of ethical values in technological development strategies and policies (Stahl 2011). Such a framework is based on three areas or nodes, seen by its proponents as ensuring a balanced mixture for collective decisions in the area of ICT: 1) regulatory framework, 2) expertise, and 3) participation of stakeholders. With the aid of an analytical map, the paper outlines actors, their interest in ageing and ICT, and the relations between each other, in these three areas.

2. THE EU INFORMATION SOCIETY POLICY WITH REGARD TO OLDER PERSONS

This section describes the instalment of a policy on Ageing and Technology in the EU, from a historical narrative perspective. For the recent and short period under consideration – i.e. 20-25 years (from the 1990s to the present day) –we see the emergence of an information society policy with regard to the elderly as part of a context that evolves over time. Three distinct periods can be observed. Table 1 summarises the main EU policy landmarks with respect to ICT and ageing.

The Big Nineties

The early nineties have been momentous historical times for Europe and the European Community/European Union (EU).[4] The Maastricht Treaty (1992-1993) marked the completion of the process of economic integration (initiated in 1986 with the Single European Act) and paved the way to the European Union. This period also observed the launching of the policy plans for the USA information superhighways, in 1992, and of Europe's information society, a year later (Mantovani/DeHert 2010). In the history of the EU, initial plans for an information society appeared in December 1993 with the European Commission's *White Paper strategy for growth and occupation* (European Commission 1993; Henten/Skouby/Falch 1996). After the White Paper, a roundtable of industrialists

---

4 | Before the Maastricht Treaty entered into force (1993) unfolding the process of European political integration (the Union), there was no 'EU' properly speaking, but the European Economic Community (EEC) or EC. Nonetheless, we refer to the EU to avoid confusion.

Table 1: Main EU Policies in the Area of ICT and Ageing (adapted from VALUE AGEING, D7.1, 2013, p.9).

	Bangemann & Blankert reports	The Lisbon Agenda	i2010: A European Information Society for Growth & Employment	The Riga Declaration	Action Plan on ICT and Ageing	Digital Agenda	European Innovation Partnership on Active & Healthy Ageing
Date	The 1990s	2000	2005	2006	2007	2010	2012
Time Horizon	n/a	2010	2010	2010		2020	2020
Description	The EU way to info society emerges as combination of economic driver and social and ethical inhibitors.	Refocused the Agenda on actions to promote economic growth and sustainable development.	Promoted the positive contribution that ICT can make to the economy and society, and emphasises ICT as a driver for inclusion and quality of life.	Promoted the use of ICT for an inclusive society.	Designed to create a significant effort in developing and deploying user-friendly ICT for the older population, and supporting policy that addresses the challenges of ageing.	Maximise the potential of ICT and define the enabling role that ICT plays in the delivery of smart, sustainable and inclusive growth. Ehealth and mhealth applications	Increase the average healthy lifespan of Europeans by 2 years by 2020, improve health and quality of life of older people, improve the sustainability and efficiency of care systems, and create growth and market opportunities for businesses.
Primary Focus	IT development and social implications of ICT	Economic Growth, Employment	Economic Growth & ICT	ICT	ICT & Ageing	ICT & Economic Growth	Ageing
Relevance to Ageing & ICT	No reference to ageing, but long-term implications of the two tier approach plus 'ethics'	ICT associated to socio-economic objectives and policies	Reference to Ageing & ICT for growth	Inclusion, digital divide, vulnerable groups	Economy also addressed in relation to ICT	Reference to Ageing	Reference to active and healthy ageing as a priority action area

released the Bangemann report, named after Martin Bangemann, commissioner of DG Enterprise and industry (Bangemann report 1994). The report underlined the potential of ICTs, a "new industrial revolution", releasing "unlimited potential for acquiring knowledge, innovation and creativity" (p. 5 and 9), and "energise every economic sector" (p.17). Endorsed by the European Council at its meeting in Corfu, the report became the 'Action Plan on Europe's Way to the Information Society' (APEWIS) (European Commission 1994). The role and the intervention of public policy were mainly oriented towards guaranteeing the conditions for a free market in ICT: "the market will drive [...]; the prime task of government is to safeguard competitive forces", states the report (p. 9).

Three years later, in 1997, the EU Commission appointed a high level group of experts to address the social aspects of the burgeoning information society (Blankert report 1997). The Blankert report warned that ICT could engender a "harmonisation by erosion of Europe's social standards" (p. 52), in particular in the field of employment and occupation. The experts took a clear stance against the expectation – which was emphatically underlined in the Bangemann report – that "knowledge, innovation, and creativity" would be flowing automatically from the "new industrial revolution". "Within an increasingly transparent information society", the group retorts, "social and regional cohesion could be undermined through a progressive erosion of the many diversified European welfare systems" (p 52).

Active Ageing, E-inclusion, and ICT for Ageing well

It is safe to say that ageing as a social phenomenon (Durkheim 1938; Searle 1995) became a matter of policy concern after the industrial revolution and the slow instalment of welfare states (de Beauvoir 1970; Laslett 1991). In the EU, ageing became a policy matter in 1999, when the Commission licenced communication 221 'Towards a Europe of All Ages'. Drafted as EU contribution to the UN International Year of Older Persons celebrated in that year, communication 221 warned that "[t]he European population will soon stop growing in size. It will then gradually start decreasing, though at different times and speeds in different countries and regions [...]. Soon our societies will have a much larger proportion of older persons and a smaller working age population" (European Commission 1999:7). The same document made a call to "prepare for longer, more active and better lives, working longer, retiring more gradually and seizing the opportunities for active contributions after retirement. "These are the best ways", continues the Commission, "to secure the maximum degree of self-reliance and self-determination through old age. This is true even in the face of fading faculties and growing dependency." After 1999, the active citizenship vision suggested by the Commission in this communication quickly gained visibility as a broader international trend: active ageing (see table 1 below). In the year 2000, the informal but influential intergovernmental forum, G8, released the Turin Charter on active ageing. The objective of the Charter was "to facilitate and support the participation of older people in economic and social life", "contribute to the goals of economic growth", and to make use "of their skills, talents and experience", including through improvements in ICT skills, "with the aim of bridging the digital divide" (paragraph I and VII, G8 Turin Charter; G8 2000).

In line with the G8, and other international bodies such as the OECD (1998) and the WHO (2002), the European Employment Strategy (EES) included as priority action the "promotion of active ageing" (EU Council 2003: Section 4. Promote development of human capital and lifelong learning). The EES, the EU policy seeking to create more and better jobs, was followed by a series of initiatives encouraging "flexicurity", accommodation of working conditions, long life training, vocational training, and flexible forms of organisation at work (European Commission 2005, 2006).

As for the information society, in 2000 the European summit in Lisbon launched the renowned plans for a "knowledge-based economy and society" (EU Council 2000: section I. paragraph 5). The Lisbon agenda envisaged an association of research and technology developments with health, education, work, and, importantly, active citizenry policies (European Commission 1999a).

After the mid-term review of the Lisbon agenda, in 2006, the inter-ministerial Declaration of Riga (Ministerial Declaration of Riga 2006) unveiled the process of *e-inclusion* (European Commission 2007). *E-inclusion* is a policy of the EU and member states designated primarily to stem the inequalities in access to technologies prevailing in groups deemed to be, for different reasons, excluded from reaping the benefits of the information society. The targeted groups included people living in rural areas, marginalised youth, migrants, persons with disabilities, and older persons.

In line with the conclusions of the Latvian summit, in 2007, the Action Plan 'Ageing well in the information society' was adopted (European Commission 2007b). The Action Plan went beyond accessibility and inclusion, targeting the concrete realities in which people age – and which could be supported by ICT: "ageing well at work"; "ageing well in the community"; and "ageing well at home" (European Commission 2007b). As a result, one could argue that, at first, ageing was connected chiefly to the employment strategy and goals of the EU; subsequently, the inclusion aspect came into play, leading to EU to focus, by means of the aforementioned action plan, on these three streams of ageing well in the information society.

ICT for Active and Healthy Ageing

The successor of the Lisbon strategy, Europe 2020, was provided with a renewed and enriched Digital Agenda (European Commission 2010). Contextual conditions had significantly changed since Lisbon. Demographic ageing is now recognised as a mega–trend alongside energy and climate change (Beblavy/Maselli/Veselkova 2014). After peaking in the nineties, a "systemic crisis" (Trichet 2010) of financial markets' economy threatens welfare states. In the EU, national governments are encouraged by the Commission to curb age-related expenditure, including health care, reform pre-retirement and pension schemes (EPC 2009, 2012; Natali/Vanhercke 2013: 253). As a consequence of the increase in life expectancy and low fertility rates, the fastest growing ageing group is the aged 80+, the oldest old (Eurostat 2010).

On the technology side, cloud computing, social networks, wearable sensors, are increasing in number and are becoming more personalised (Hood/Flores 2012). Industry has acquired the capacity to purposely develop or adapt existing products to cater for the identified needs of users, to reduce hospital stays and the degree of surgical intervention, and to enable self-care at home (Llewellyn/Chaix-Viros 2008). Against this backdrop, at the end of 2011, a European Innovation Partnership on Active and Healthy Ageing (EIP on AHA) was launched (European Commission 2011).

The European Innovation Partnership on Active and Healthy Ageing (EIP on AHA)

Part of Europe 2020, European Innovation Partnerships (EIPs) aim at disseminating innovation through better framework conditions for commercialisation of new solutions in the areas of raw materials, water, smart cities, agriculture, and ageing. Specifically, the objective of the EIP on Active and Healthy Ageing (AHA) is to improve the quality of life of older people: its motto and broad political vision is "to add two healthy life years to the average healthy life span of European citizens" by 2020.[5]

The EIP on AHA develops along two main avenues, reference sites and cooperation on thematic areas. Reference sites are regions, cities, or integrated hospitals/care organisations that implement a comprehensive, innovation based approach to active and healthy ageing and can give concrete evidence and illustrations of their impact on the ground.[6] The second avenue, cooperation on thematic areas, is organised around six action groups delving on, e.g., prescription and adherence to treatment, personalised health management, fall prevention, prevention of functional decline and frailty, etc.

A High Level Steering Group coordinates the EIP on AHA. It includes the European Commission (DG responsible for information society and health and consumer rights), national and regional authorities, industry, professionals, elderly and patient organisations, and other interest groups (European Commission 2011).

As submitted in the first expert assessment of the EIPs, "EIPs should be leveraged as 'outriders', first movers, able to test ideas, generate feedback, and scale opportunities to address societal challenges" (European Commission 2014: 9). In other words, the key success indicator of the EIP is, rather than the number of actions or activities adopted, the "capacity to drive large scale change" (European Commission 2014: 9).

The EIP on AHA marks a shift in the EU policy on ICT and ageing. At first focused on active ageing at work; after 2007 portrayed as a hopeful promise to support living independently at home, at work, and in the community (e-inclusion); with the 2012 EIP on AHA, the EU policy on ageing and technology embraces health. The association between health and ageing is clearly expressed in its resounding objective, "to add two healthy life years to the average healthy life span of European citizens". In addition, the partnership emphasises the role of national, regional and local stakeholders, which are seen as best placed to demonstrate the benefits of ICT for active and healthy ageing.

---

5 | Research indicates that the objective is out of reach due to persistent inequalities in access to health care across Member States (Jagger et al. 2013).

6 | As of July 2013, out of 56 candidate sites that applied, 32 were awarded the title of 'European Reference Site'.

Normative Framings of Ageing and Technology

In the historical review we have encountered expressions such as "ageing well", "active ageing", "active and healthy ageing", in the Action Plan of 2007, in the G8 meeting of 2000, and in the EIP on AHA in 2011, respectively. In order to enrich the understanding of the policy strategy of the EU, this sub-section provides a list of different framings of ageing developed by scholars in gerontology, here presented not in chronological order, but in a manner so as to highlight their contrasting characteristics. This plurality of meanings attributed to the elderly suggests that old age itself does not carry a fixed meaning, but that alternative meanings have been coined. Even if the role of technology is explicitly mentioned only under "ageing well", all definitions can be seen as framings of ageing and technology (A&T). We here discuss these framings individually and provide a summary in Table 2.

Successful Ageing

The concept of successful ageing or "optimal ageing" is credited to Baltes & Baltes (1990), who claimed that success or optimal ageing requires a mixture of "selection, optimisation and compensation" (Baltes/Baltes 1990: 2). The example they gave is that of a famous pianist who, getting older, selects what to play, rehearses selectively, and plays "tricks" to compensate declining dexterity (Baltes 1997). In 1997, Rowe and Kahn provided a definition of successful ageing premised on the retention of functional capacities (Rowe/Kahn 1997: 433). This definition has been criticised for placing too much emphasis on the physical, functional health of people, as a completely disease-free older age is unrealistic for most individuals (Bowling/Dieppe 2005).

Productive Ageing

Coined in 1985 by American gerontologists Butler and Gleason (1985), productive ageing draws the attention to the contribution and the value of older persons in society and in the economy. Within the frame of this definition, both positive connotations of continuous engagement and more functionalist visions of later life are represented. More recent research in this area by Kaye, Butler & Webster (2003) has recognised sense of purpose in work activities in later life as central.

Active Ageing

Sponsored in international forums and organisations such as the G8 (2000), mentioned earlier, developed and systematised in seven principles by the English gerontologist Alan Walker (Walker 2002, 2009), active ageing was introduced first as an incentive for adults to remain in the workforce for longer (European Commission 2005a). It was subsequently adapted to convey a more inclusive message, placing the rights of older persons at its centre (Moulaert/Paris 2013). As of the end of 2014, an Active Ageing Index provides a set of benchmarks to measure 'the untapped potential' of seniors across the 27 EU Member States and beyond (Karpinska/Dykstra 2014: 2).

Healthy Ageing

The term "healthy ageing" was promoted by the World Health Organisation as early as 1980 (Davey/Glasgow 2006). Originally informed by medical views on the incidence of chronic diseases and physical functional capacities, later definitions underline behavioural and social factors of healthy ageing (Kaplan/Strawbridge 1994). One of the definitions refers to it as "the increase the number of healthy life-years (HLYs) lived without or with minimal functional limitation, disability or disease" (Robine/Mathers/Bone/Romieu 1993: 13). This framing has gained importance further to advancements in medical health technologies.

Active and Healthy Ageing

Arguably a decisive impulse to the emergence of the term "active and healthy ageing" was given by bio- and health gerontology studies (Fried/et al. 2001; Lafontaine 2009). This field of research investigates the physical mechanisms, notably the diseases that trigger physical changes in old age (Masoro 2006). A contemporary biogerontologist from the UK, Aubrey de Grey, is so convinced that ageing should be seen as a disease that he researches the possibility to "address ageing just as effectively as we address many diseases today" (De Grey 2006:66). This avenue of research has met with the criticism of Roger Scruton, de Grey's countryman and historian. Scruton retorts that to equate old age to a disease is tantamount to turning all healthy people in unhealthy subjects by default. The next step is the creation of a health "thought police": "If you pursue a life of risk-taking and defiance the thought-police will track you down. For a population of docile and loveless geriatrics is the telos of the welfare state", he scorns (Scruton 2012: 434). These diverging views spring to mind reading the motto or objective of the EIP on AHA, mentioned earlier. In essence, the partnership on active and healthy ageing seeks to maximise opportunities for health in old age (European Commission 2011: 11-13). The goal "to add two healthy life years to the average healthy life span of European citizens" (that we could associate with de Grey's view of old age as a disease) taunts the question suggested by Scruton: if it is possible to add two healthy life years to the average healthy life span of European citizens, why should we do it?

Ageing Well

Over the last few decades, there has been a great increase in publications, particularly in popular media, magazines, books, television, etc. over what constitutes a good old life. In the academic literature, the term ageing well evokes social models of quality of life (QoL). Bowling (2005) describes them as investigating the positive characters of what compounds to a good life: morale, satisfaction, happiness, well being, pleasure, social well being. These topics do not resonate in the EU policy discourse on ageing well (Commission of the European Communities, 2007a), which is less focussed on well-being, pleasure, happiness and satisfaction than it is on the visions of an active and productive old age at work, in the community, and at home. However, the Action Plan on ageing well devotes attention to the societal and ethical implications of using ICT to age well. In this document, we find a caveat stating that "when ethical concerns are ignored or not fully taken into account by the technology developed, they lead to a rejection by the older person and his informal carers and then constitute a barrier to market uptake" (European Commission 2007a, p. 7). The same document advises that users should also have the right "to overrule or switch off the technology" and "to opt out completely from using the services, should they so wish". Such rights", it continues, "must be built into the services" (European Commission 2007a, para. 5.3.4.).

As anticipated, the plurality of framings suggests that old age itself does not carry a fixed meaning, but that alternative meanings have been coined. Out of this plurality, the "active ageing" narrative echoes in particular with the EU social and economic policy agenda on ageing societies. For our research on ageing and technology, the "ageing well" definition can be seen as an early framing. In it, we find the idea that using technology to age well must coexist with the recognition that some may opt out completely, and that they should be free to do so. After all, using assistive technologies services, to mention one application of ICT for ageing, means delegating at least a part of the process of care to a machine or to a third persons who is not present. Some may prefer not to.

The mediating, humble tone that we register in 2007 wanes in the other framing that is immediately relevant for our discussion. The objective that, in the year 2012, crowns the EIP on Active and healthy ageing takes for granted only the positive narratives of living longer lives active and healthy with ICT. Unlike ageing well, the active and healthy ageing is, as such, potentially divisive. The pace of technological advancements; the increasing number of older persons; the social and economic questions posed by demographic ageing (e.g., on resource allocation, justice between generations, care, etc.); the elevation of narratives of non conformity to the norm of ageing and deterioration, all these factors suggest that the spat between de Grey and Scruton construed above may be not too far off the tables of decision makers.

Table 2: Framings of Ageing – an Overview

FRAMING	DESCRIPTION
Successful Ageing	A mixture of selection, optimisation and compensation skills (Baltes/Baltes 1990)
	Low probability of disease and disease-related disability, high cognitive and physical functional capacity, and active engagement with life (Rowe /Kahn 1997)
Productive Ageing	The ability of older individuals to take part in paid workforce or volunteer activities, to maintain the family and themselves as independently as possible, without assistance from others (Butler/Gleason 1985)
	Any activity by an older individual that produces goods and services, or develops the capacity to produce them, whether they are paid for or not (Bass/Caro/Chen 1993)
Active Ageing	A political concept which seeks to optimise opportunities for older adults regarding their participation in society according to their desires and capabilities (Walker 2002; 2009)
Healthy Ageing	The increase the number of healthy life-years (HLYs) lived without or with minimal functional limitation, disability or disease (Kaplan/Strawbridge 1994)
Active and Healthy Ageing	To maximise opportunities for health, participation and security in order to enhance quality of life in an ageing population (European Commission 2011)
Ageing well	Ageing well at work or active ageing at work: staying active and productive for longer, with better quality of work and work-life balance with the help of easy-to-access ICT, innovative practices for adaptable, flexible workplaces, ICT skills and competencies and ICT enhanced learning (resp. e-skills and e-learning).
	Ageing well in the community: staying socially active and creative, through ICT solutions for social networking, as well as access to public and commercial services, thus improving quality of life and reducing social isolation (one of the main problems of older people in rural, scarcely populated areas, as well as urban areas with limited family support)
	Ageing well at home: enjoying a healthier and higher quality of daily life for longer, assisted by technology, while maintaining a high degree of independence, autonomy and dignity (European Commission 2007a, 2007b)

3. ETHICS IN THE EUROPEAN GOVERNANCE OF AGEING AND TECHNOLOGY: A MAPPING EXERCISE

In a seminal article of 1994, American gerontologist Harry Moody depicted "four scenarios for an aging society":[7] in the four scenarios, societies make choices concerning technology developments to invest in medical treatments to reimburse, eligibility and access to regenerative medicine, the limits of individual autonomy, as in cases of planned death decisions, etc. These choices, Moody warns, do not appeal only to procedural theories of justice or to bureaucratic decisions about the allocation of resources. Choices like these presuppose specific ideas about the meaning of later life: "Certain ideas about meaning and value – for example quality of life, successful ageing, or intergenerational solidarity", states Moody, "are intrinsically problematic because they involve difficult philosophical questions about the purpose of human life" (Moody 1994: 59). "Public policy", he contends, "can and must take seriously a variety of different ideas about a good old age" (Moody 1994, ibid.). Taking seriously a variety of different ideas about a good old age is a sensible recommendation, also in the area of A&T. As already pointed out by Winner in 1980, "artefacts have politics", which means that technological devices and systems contain possibilities for different ways of ordering human activities, of exclusion or inclusion (Winner 1980, p.134).[8]

---

7 | The first scenario, prolongation of morbidity, relies on medicine and scientific developments to enable older persons to live longer as unencumbered as possible by old age diseases. The second scenario, compression of morbidity, presupposes an ongoing and carefully designed programme of health control to postpone any disease or chronic illnesses, until just before death. The third scenario, prolongevity, presupposes investments in the genetic potential of higher maximum life span; the fourth scenario is based on the assumption that the meaning of old age lies with the finitude of life, accepted not just as an individual choice, but as 'a matter of collective policy'.

8 | One of the many examples offered relates to the bridges built over the parkways linking Long Island and New York. Winner remarks that many overpasses are extraordinarily low. The reason for this is that they were built to specifications that would discourage the presence of buses, a means of transport normally used by poor people and Afro-Americans. This is an example of social-class bias and race-based exclusion embodied in artefacts.

Starting from these premises, the second part of this contribution looks at European governance on A&T, narrowly defined as the actor configurations and procedures available in the EU to expose and address ethical questions and views on A&T. To identify the procedures available in the EU where ethics of ICT and ageing can be exposed and addressed, the concept of the EU-ETICA project is used here. This project asked how values are considered in research and technology developments policy (Stahl 2011; ETICA 2010). In particular, ETICA highlighted the role of policy in creating the "infrastructure for the development of responsibility" (Stahl 2011: 152). Such a framework, according to Stahl and his colleagues, needs to be provided and must cover at least the three main areas of policy activity: 1) regulatory framework, 2) ethics observatory, and 3) stakeholders' involvement and participation (Stahl 2011; ETICA 2010[9]).

In our interpretation of ETICA's recommendations in light of the specific issues at hand (Value Ageing 2013), the *regulatory framework* encompasses the international and European bodies setting rules, technical codes, standards, etc., that contribute to set the conditions for ICT for Ageing to develop and be deployed in living settings. The recommendation on '*ethics observatory*' points at the role of experts. The role of knowledgeable and recognised experts is increasingly important in a context of institutionalised social roles and structures, specialisation of knowledge and rapid techno-scientific developments. In the area of research concerned with ethical issues in ICT for ageing, expert papers and position statements provide guideposts for collective reflection and debate around ethical dilemmas. *Stakeholder involvement* refers to civil society and other stakeholders prepared to engage on a content level with the policy community as well as with the technical community (Stahl 2011). In the area of ageing and ICT, forums for stakeholders' involvement are important to reflect the concerns and expectations that the elderly have towards technological solutions purposefully developed for their benefit. The following section contemplates how stakeholders in the sphere of the EU concur to shape the direction of a European policy on ageing and technology. In the first section, we have referred almost exclusively to acts of the Commission.

---

9 | The ETICA project was a research project on "Ethical Issues of Emerging ICT Applications" funded by the European Commission under the 7th Framework Programme (GA 230318). It ran from April 2009 to May 2011. ETICA's main objective was to identify ethical issues of emerging technologies and their potential application areas in order to analyse and evaluate ethical issues arising from these. See http://www.etica-project.eu.

Mapping the Actors and Networks relevant to A&T Policy in the EU

The following map presents an analytical map of the main stakeholders involved in the process of policy development in relation to ICT for ageing, at EU level. The field denotes a relational space that encompasses private and public, European, and international actors and networks. As for any schematic representation, the map does not seek to be exhaustive.

The map is composed of the following main elements:

- The policy-making environment, sub-divided into areas, accommodates for the specialisation of actors according to function and/or interest;
- Policy actors (individual entities or compounded groups), represented by circles.
- Formal and/or informal networks, represented by hexagons, in which actors are involved to pursue interests and contribute to policy formulation. These can also be formal sittings where multiple views are represented, e.g. in the case of the European Parliament (EP).
- Each actor is linked to one or more networks. Solid lines represent links to working groups or departments, established by/in the central node. Dashed lines represent the most relevant formal and informal interactions. These links are indicative of what has been identified as major ties, rather than exhaustive. The filled arrows (e.g. from industry to other areas) represent pervasive forms of influence that cannot be mapped within this scope.

In the following, we discuss selected elements of the governance space of ICT and ageing that are of particular relevance to the incorporation of ethical concerns, with specific reference to the analytical categories offered by Stahl (2011).[10]

---

[10] | Deliverable D7.1 of the VALUE AGEING project offers a full description of the individual actors represented in this map (VALUE AGEING 2013).

Figure 1: Actors and Networks involved in A&T Policy in the EU (adapted from VALUE AGEING, 2013: 24)

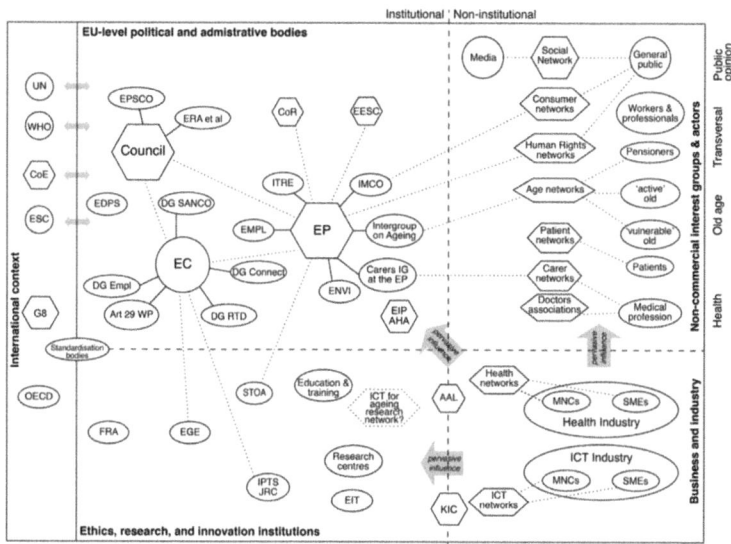

List of acronyms for Figure 1:
AAL: Ambient Assisted Living
CoE: Council of Europe
CoR: Committee of the Regions
DG CONNECT: Directorate General for Communications Networks, Content and Technology
DG EMPLO: Directorate General for Employment and Social Affairs
DG RTD: Directorate General for Research and Innovation
DG SANCO: Directorate General for Health and Consumers.
EC: European Commission
EDPS: European Data Protection Supervisor
EESC: European Economic and Social Committee
EGE: European Group on Ethics in Science and New Technologies
EIP AHA: European Innovation Partnership on Active and Healthy Ageing
EIT: European Institute of Innovation and Technology
EMPL: Committee on Employment and Social Affairs
ENVI: Committee on the Environment, Public Health and Food Safety.
EP: European Parliament
EPSCO: Employment, Social Policy, Health and Consumer Affairs Council
EIP on AHA: European Innovation Partnership on Active and Healthy Ageing
ERA: European Research Area
ESC: European Social Charter
FRA: European Agency for Fundamental Rights
G8: Group of Eight
IMCO: Committee on Internal Market and Consumer Protection
IPTS-JRC: Institute for Prospective Technological Studies Joint Research Centre
ITRE: Committee on Industry, Research and Energy
KIC: Knowledge and Innovation Community
MNCs: Multinational Corporations
OECD: Organisation for Economic Co-operation and Development
SMEs: Small and Medium Enterprises
STOA: Science and Technology Options Assessment
UN: United Nations
WHO: World Health Organization

Regulation, Expertise and Stakeholder Participation in the European Governance of Ageing and Technology

Having mapped the actors and their respective networks, we now turn to discuss the most important actors, their interest in ageing and ICT, and their relation between each other in the three areas of governance mentioned in the previous section.

Regulatory Capacity

As discussed earlier, active ageing originates in the international context of the G8 of 2000, which licensed the 'Turin Charter: Towards Active Ageing' (G8 2000). DG EMPLO took an active part in the Turin meeting. A dashed line in the map represents this relevant, informal, tie. Another dashed line links the EU Commission, notably DG SANCO, to the World Health Organisation (WHO) and its framework on active ageing (WHO 2002). A filled arrow, representing a pervasive form of influence that cannot easily be reduced to inter-agencies communication, connects the United Nations (UN) and the EU areas. As mentioned earlier, active ageing combines a "live longer-work longer" agenda with a more inclusive message, based on equality in access to ICT. In particular, the 2002 United Nations Madrid Action Plan (MIPAA) and the UN Economic Council of Europe (UNECE)'s Regional Implementation Strategy (MIPAA/RIS 2002) emphasised the need for fair and equitable access to technology.

Formally, the EU agenda on A&T policy is based on action plans (2007), joint initiatives (AAL), and partnerships (the EIP on AHA) (Value Ageing 2013; Mantovani/De Hert 2010). The Commission, through its Directorates-General (DGs), plays a major role in setting the agenda for these initiatives, as these pertain to the Union's shared competence in research and technology developments (Articles 179-190, TFEU 2009). In this field, the Commission has the power to launch policy initiatives in agreement with the Council of the European Union (EU Council). In December 2012, the Employment, Social Policy, Health and Consumer Affairs Council (EPSCO), adopted a resolution backing the Commission agenda on active ageing and endorsed a list of nineteen 'Guiding Principles for Active Ageing and Solidarity between Generations', a checklist for national authorities and other stakeholders on how to promote active ageing (Council of the European Union 2012).

The European Parliament (EP) plays an important co-decision role in A&T policy. A number of European Parliament committees are particularly active in this area: the ITRE Committee (Committee on Industry, Research and Energy) followed the Ambient Assisted Living (AAL) initiative and dealt with information exclusion in the past; the ENVI Committee (Committee on the Environment, Public Health and Food Safety) was in charge of the EIP on AHA of behalf of the EP. Outside the formal decision making process, the EP includes 'inter-groups', i.e., informal forums for MEPs and civil society organisations (CSOs) on crosscutting themes and specific issues. In the field of ageing, important forums are the Intergroup on Ageing and Intergenerational Solidarity and the Carers Interest Group.

In addition, two advisory bodies, the European Economic and Social Council (EESC) and the Committee of the Regions (CoR) are active in outreach activities that involve regional and local authorities, the latter, and in providing advice and input from socio-economic operators and local authorities, the former (EESC 2012). Recently, in partnership with AGE Platform Europe (see below), the CoR played an active role in the activities organised during the 2012, the 'European Year for Active Ageing and Solidarity between Generations.'

Expertise

The European Group on Ethics in Science and New Technologies (EGE), the former Group of Advisers on the Ethical Implications of Biotechnology (GAEIB), is established under the presidency of the Commission (Plomer 2008). EGE's mandate is broadly defined as to cover all areas of the application of science and technology, including ethical oversight in the allocation of funding to research activities under the Framework 7 programme and under H2020 (Commission Decision 2005/383/EC and 2010/1/). When research and technology developments raise ethical concerns, the European Commission can, at its discretion, ask for an opinion of the fifteen experts group; EGE can also act on its own initiative issuing 'Opinions towards the Commission'. In any case, opinions have the status of non-binding ethical advice.

EGE has not issued any comprehensive opinion covering A&T. In its Opinion 26 of 2012 on the Ethics of Information and Communication Technologies, EGE underlined that "older people tend to face other obstacles such as cost, skills, disability access and attitude, as well as lack of awareness and understanding" (European Commission 2012: 50). There are other reports that are relevant to the elderly in the information society: in addition to Opinion 26, Opinion 20 of 16 March 2005 on ethical aspects of ICT Implants in the Human Body is important (European Commission 2012).

A relation similar to that linking EGE with the European Commission exists between STOA, the Science and Technology Options Assessment office, and the European Parliament. STOA's task is to provide MEPs with advice and information on technology development (European Parliament, 2012) and its research is often carried out in partnership with external experts (European Parliament, n.d.). The work of STOA is, however, less influential as compared to that of EGE in terms of the number of reports compiled, activities organised (e.g. seminars, consultations, newsletters), contacts with national ethical bodies, and visibility.

Since 2007, when the "ageing well" action plan was adopted, the EU commission has funded research about the ethical, social and psychological implications of technologies for ageing (SENIOR project, 2008). The primary objective of these projects has been the creation of appropriate framework conditions conducive to the uptake of ICT for ageing. "When ethical concerns are not addressed properly", warned the Commission in the 2007 Ageing Well in the Information Society Action Plan, "they lead to a rejection or low uptake of technology solutions" (European Commission 2007a: 46 and 2007b). Among its activities, the Institute for Prospective Technological Studies Joint Research Centre (IPTS-JRC) in Seville focused on the economics of ICT for long-term care and on training and educations in e-skills (Cabrera/Malanowski 2009). Similarly, the European Institute of Innovation and Technology (EIT), through its Knowledge and Innovation Community (KIC) on innovation for healthy living and active ageing (created in 2014), aims at fostering cooperation between research centres, industries, and private and public actors interested in adopting technological solutions for ageing societies.

Participation of Stakeholders

The EU, and in particular the European Commission, has a long track record of practice in involving a wide range of stakeholders to discuss key issues and to share ideas. EU officials, in particular Commission officials, meet routinely with representatives from industry, trade unions and academia trying to agree on a joint course of action. The EU's relations with stakeholders is often criticised for who gets included and who does not, and who sets the agenda (Kutay 2014; Cooke/Kothari 2002; McLean 2011).

Our research indicates that business and industry have a prominent role in networks such as Ambient Assisted Living (AAL) and the European Institute of Innovation and Technology (EIT) Knowledge and Innovation Community (KIC), mentioned earlier. Large industry representatives also appear to be significantly involved in drafting EU level policies (Europa 2013).

'Non-commercial interest groups and actors' include groups corresponding to generic categories, such as consumer networks, age networks, patient networks, carer networks, and professional associations. For reasons of space, we can only point to the marginal role of older persons' associations and carers in the field of A&T. To our knowledge, AGE Platform Europe[11] is the only organisation that plays an active role in the context of A&T. AGE Secretariat is part of the high level steering committee of the EIP on AHA, and its members are consulted and reactive in issuing opinions and providing advice to the Commission in areas such as accessibility and remote monitoring systems for long-term care.

4. CONCLUSION – *QUESTA E QUELLA*. THE NEED OF MORE ATTENTION TO SUBSIDIARITY?

This chapter originated in a research project funded by the EU, VALUE AGEING, in charge of studying the incorporation of European fundamental values in ICT for older persons or ageing and technology. The chapter has provided a historical review of EU policies on A&T, and directed attention to the main normative framings of ageing, their diversity and main advocates. These reviews led to a consideration of the actors and networks contributing to shape EU policies in this area – literally through the elaboration of a "map". The mapping exercise put into perspective three areas of EU governance as providing different avenues for collective decisions in the area of technology developments and research: regulatory capacity, expertise, and participation of stakeholders. We expect this map to provide useful reference and tool for the positioning of decision-making in this area.

The chapter has showed that the European Union is active in developing policies that attempt to combine technology, innovation and new ways of cooperation in the context of ageing societies. EU actors, committees and projects are also active in addressing the ethical questions and considerations that emerge from the encounter of technology and ageing. As the regulatory

---

11 | As stated on their web-site, AGE Platform Europe is a European network of more than 150 organisations of and for people aged 50+ representing directly over 40 million older people in Europe. http://www.age-platform.eu/about-age.

and expertise sections suggest, the EU Commission currently plays a central role in A&T policy; it also seems to play a large role in framing the questions that should or should not attract ethical consideration.

One of the questions that arise at the end of this contribution wholly centred on the EU level is whether ethical considerations in A&T –the question could be extended to other areas where the EU intervenes such as biotechnology (Tallacchini 2009) – pertain to the supranational domain. This would justify the pro-activeness of the Commission over member states and local communities. For the European Group on Ethics and New Technologies (EGE), mentioned earlier, cultural differences can coexist with a common core of fundamental values, enshrined in the EU Charter of Fundamental rights. According to others, for instance, Tallacchini, "ethics and cultural values are regulated on the national level and follow the principle of subsidiarity" (Tallacchini 2009, p.288). The national level and the principle of subsidiarity have been beyond the scope of this chapter – which has chiefly dealt with actors and processes involved at European level. There is an opportunity for further research in this direction. Indeed, there is arguably an urgent need to hear and learn from practices and experiences from the local level.

The analytical review offered in this contribution suggests that old age is adroitly construed as a risk. The consequences and associated uncertainties of ageing, so to speak, come with the suggestion that there are boundaries to an active life, which are seen as a risk to be overcome or mitigated. The eventuality of ageing – which is, in fact, a certainty, but more a matter of timing and qualification – is painted in dim colours that, we are being told, technology could help address through more control (e.g. via lifestyle changes) and precautionary measures. But, if advances in technology and science are portrayed as a hopeful promise, they do not automatically entail commensurate gains in the effective policy responses to complex social, economic and health problems which also occur in ageing (Jasanoff 2009). In this contribution, regulatory capacity, expertise, and participation of stakeholders have been introduced as the three legs of a governance approach that, far from being a solution, draw attention to the fact that framings and policy decisions on technology developments, however benevolently introduced, presuppose different ideas of later life – that should themselves be open for discussion and co-construction (particularly in their associated normative aspects). The coexistence of regulatory capacity, expertise and participation provides, in theory, a space for recursively (re-)considering and (re-)negotiating ethically-informed decisions in a domain characterised by the double uncertainties of prospective technological change (ICT) and emerging socio-cultural values (related to ageing societies). In practice, however, there are limitations inherent to the approach put forward by the EU that invites us to mobilise the national, regional, and local levels (subsidiarity).

One relevant limit of the current EU approach is linked to the representation of older persons. Representation quite literally refers to play on stage, to be an actor playing a role. Those who represent, e.g., interest groups, parliamentary groups, etc., may hold different views than those whom they represent. While official representatives are powerful players with privileged access to forums of decision-making and contestation, many social actors lack this access and power. Out of these social actors, furthermore, the frail or the poor are the least represented because they currently exercise very little political clout and agency. Citing the example of nursing homes in the United States during the 1950s, anthropologist Athena McLean (2011) reminds us that in the past policy solutions for ageing problems "have often been driven by the interests of the most powerful stakeholders over the wishes of the vulnerable persons who are most directly affected by the solutions" (McLean 2011: 323). "Today", she continues, "the e-solution that has been embraced to promote digital access and independent living of older persons has similarly been crafted mainly by stakeholders in positions of power" (McLean 2011: 323). In order to mend the representation asymmetry, the EU policy on ageing and technology should open the door to practices and experiences springing from the interactions of laymen elderly with technology. The reference sites and the action groups promoted by the EIP on AHA can be saluted as a positive development in the right direction. It remains to be seen whether these experiences will be able to critically address the legitimacy and ethical acceptability of political objectives of today's EU active and healthy ageing agenda.

REFERENCES

Baltes, P.B. (1997): "On the incomplete architecture of human ontogeny: Selection, optimization, and compensation as foundation of developmental theory." In: American Psychologist 52, pp. 366-380.

Baltes, P.B./Baltes, M.M. (1990): "Psychological Perspectives on Successful Aging: The Model of Selective Optimization with Compensation." In Baltes, P.B./Baltes, M.M. (eds.), Successful Aging: Perspectives from the Behavioral Sciences, New York: The European Science Foundation, pp.1-34.

Bangemann Report (1994): Europe and the Global Information Society, http://www.cyber-rights.org/documents/bangemann.htm#chap1 (last access: 01.08. 2014).

Beblavy, M./Maselli, I./Veselkova, M. (2014): Let's get to work. The future of Labour in Europe, Brussels: CEPS.

Blankert Report (1997): "Building the European Information Society for Us All", Luxembourg: Office for Official Publications of the European Communities, http://www.epractice.eu/files/media/media_688.pdf (last access: 05.11.2009).

Bowling, A. (2005): Ageing well: Quality of life in old age, Maidenhead (UK): Open University Press.

Bowling, A./Dieppe, P. (2005): "What is Successful Ageing and Who Should Define It?" In: British Medical Journal 7531/331, pp.1548-1551.

Butler, R. N./Gleason, H.P. (1985): Productive aging: Enhancing vitality in later life. New York: Springer.

Cabrera, M./Malanowski, N. (eds.) (2009): Information and communication technologies for active ageing: Opportunities and challenges for the European union (Vol. 23 of Assistive Technology Research Series), Amsterdam: IOS Press.

Cooke, B./Kothari, U. (2002): Participation: The New Tyranny?, London: Zed Books.

Council of the European Union (2012): Council Declarations on the European Year for Active Ageing and Solidarity between Generations: The Way Forward Brussels: Council of the European Union http://register.consilium.europa.eu/pdf/en/12/st17/st17468.en12.pdf (last access: 13.02.2014).

Davey, J./Glasgow, K. (2006): "Positive Ageing – A Critical Analysis." In: Policy Quarterly 4/2, pp. 21-27.

De Beauvoir, S. (1970): La vieillesse, Paris: Gallimard.

De Grey, A. (2006): "We will be able to live to 1000." In Moody, H.R. (ed.), Aging: concepts and controversies, London: Sage, pp. 66–68.

Durkheim, E. (1938): The rules of sociological method, New York: The Free Press.

Economic Policy Committee (EPC – Ageing Working Group) (2009): 2009 Ageing Report: Economic and budgetary projections for the EU-27 Member States (2008-2060). http://ec.europa.eu/economy_finance/publications/publication16034_en.pdf (last access: 12.07. 2013).

Economic Policy Committee (EPC – Ageing Working Group) (2012): The 2012 Ageing Report: economic and budgetary projections for the 27 EU

Member States (2010-2060). http://europa.eu/epc/pdf/2012_ageing_report_en.pdf (last access: 12.07.2015)
EIP on AHA (2011), Action Plan on 'Innovation for Age-friendly buildings, cities & environments', https://ec.europa.eu/research/innovation-union/pdf/active-healthy-ageing/d4_action_plan.pdf (last access: 14.07. 2013).
ETICA (2010): Deliverable D4.2: Governance Recommendations, The ETICA project, EU 7th Framework Programme (GA 230318), http://www.etica-project.eu/deliverable-files (last access: 7.11.2014).
EU Council (2000): Presidency conclusions, Lisbon European Council, 23-24 March 2000, http://www.europarl.europa.eu/summits/lis1_en.htm#ann (last access: 31.01.2015).
EU Council (2003): Guidelines for the employment policies of the Member States, Brussels: OJ L 197.
Europa (2013): Digital Agenda for Europe: A Europe 2020 Initiative, https://ec.europa.eu/digital-agenda/en/participants-list-6 (last access: 28.02.2013).
European Commission (2005): Decision 2005/383/EC on the renewal of the mandate of the European group on ethics in science and new technologies. Brussels: OJ L 284.
European Commission (2010): Decision 2010/1/ on the renewal of the mandate of the European Group on Ethics in Science and New Technologies. Brussels: OJ L 1, 5.1.2010.
European Commission (1993): White Paper on Growth, Competitiveness, Employment: The Challenges and Ways Forward into the 21st Century, Brussels: COM (93) 700 final/A and B.
European Commission (1994): Action Plan on the Europe's Way to the Information Society (APEWIS), Brussels: COM (94)347.
European Commission (1999): Towards a Europe for all ages-Promoting prosperity and intergenerational security, Brussles: COM(1999) 221 final.
European Commission (1999a): eEurope–An information society for all, Brussels: COM(1999) 687 final.
European Commission (2005): Communication to the Spring European Council, Working Together for Growth and Jobs: A New Start for the Lisbon Strategy. Communication from President Barroso in Agreement with Vice-President Verheugen, Brussels: COM/2005/0024 final.
European Commission (2005a): 'i2010–A European Information Society for growth and employment, Brussels: COM(2005) 229 final.
European Commission (2006): Time to move up a gear: The new partnership for growth and jobs, Brussels: COM(2006) 30.

European Commission (2007): To be part of the Information Society, Brussels: COM (2007) 694 final.
European Commission (2007a): Ageing Well in the Information Society. An i2010 Initiative: Action Plan on Information and Communication Technologies and Ageing, Brussels: SEC/2007/0811 final.
European Commission (2007b): Ageing well in the Information Society: Action Plan on Information and Communication Technologies and Ageing. An i2010 Initiative, Brussels: COM (2007) 332 final.
European Commission (2010): A Digital Agenda for Europe, Brussels: COM/2010/0245 f/2.
European Commission (2011): Operational Plan. Strategic Implementation Plan for the European Innovation Partnership on Active and Healthy Ageing: Steering Group Working Document, Brussels: European Innovation Partnership, http://ec.europa.eu/research/innovation-union/pdf/active-healthy-ageing/steering-group/operational_plan.pdf#view=fit&pagemode=none (last access: 15.9.14).
European Commission (2012): Ethics of Information and Communication Technologies: Opinion No 26. Opinion of the European Group on Ethics in Science and New Technologies to the European Commission, http://ec.europa.eu/bepa/european-group-ethics/docs/publications/ict_final_22_february-adopted.pdf (last access: 14.3.2014).
European Commission (2014): Outriders for European Competitiveness European Innovation Partnerships (EIPs) as a Tool for Systemic Change. Report of the Independent Expert Group, http://ec.europa.eu/research/innovation-union/pdf/outriders_for_european_competitiveness_eip.pdf (last access: 05.01. 2015).
European Economic and Social Council (EESC) (2012): Opinion on 'Horizon 2020: Road maps for ageing', Brussels: OJ C 229, pp. 13–17.
European Parliament (2012): Science and Technology Options Assessment: Annual Report 2011, Brussels: European Communities.
European Parliament (n.d.): European Parliament/Science and Technology Options Assessment: About STOA, http://www.europarl.europa.eu/stoa/cms/home/about (last access: 13.04.2014).
Eurostat (2010): Work session on demographic projections, Lisbon, 28-30 April 2010, edition Methodologies and Working papers, Luxembourg: Publications Office of the European Union, http://ec.europa.eu/eurostat/statistics-explained/index.php/Population_structure_and_ageing (last access: 01. 04. 2015)

Fried, L.P./Tangen, C.M./Walston, J./Newman, A.B./Hirsch, C./Gottdiener, J./Seeman,T./Russell, T./Kop, W.J./Burke, G./McBurnie, M-A. (2001): "Frailty in older adults: evidence for a phenotype." In: Journal of Gerontology: Medical Sciences 56/3, pp.146–156.

G8 (2000): Labor Ministers Conference, Turin Italy, November 10-11, 2000. The 'Turin Charter: Towards Active Aging', http://www.g7.utoronto.ca/employment/labour2000_ageing.htm (last access: 01. 08. 2014).

Henten, A./Skouby, K.E./Falch, M. (1996): European planning for an information society. Telematics and Informatics, 13/2, pp. 177-190.

Hood, L./Flores, M. (2012): "A personal view on systems medicine and the emergence of proactive P4 medicine: predictive, preventive, personalized and participatory." In: New biotechnology, 29/6, pp. 613-624.

Jagger, C./McKee, M./Christensen, K./Lagiewka, K./Nusselder, W./Van Oyen, H./Cambois, E./Jeune, B./Robine, J.M. (2013): "Mind the gap—reaching the European target of a 2-year increase in healthy life years in the next decade." In: The European Journal of Public Health 23/5, pp. 829-833.

Jasanoff, S. (2009): "The past as prologue in life extension." In: Society. 46/3, pp. 232-234.

Kaye, L.W./Butler, S.S./Webster, N.M. (2003): "Toward a Productive Ageing Paradigm for Geriatric Practice." In: Ageing International 28/2, pp. 200-213.

Kaplan, G.A./Strawbridge, W.J. (1994): "Behavioral and Social Factors in Healthy Aging." In: Abeles, R.P./Gift, H.C./Ory, M.G. (eds.), Aging and Quality of Life, New York: Springer, pp. 57-78.

Karpinska, K./Dykstra, P. (2014): The Active Ageing Index and its extension to the regional level. Discussion paper. Peer Review on the Active Ageing Index, Luxembourg: Publications Office of the European Union..

Kutay, A. (2014): Governance and European Civil Society: Governmentality, Discourse and NGOs. New York: Routledge.

Lafontaine, C. (2009): "The postmortal condition: from the biomedical deconstruction of death to the extension of longevity." In: Science as Culture 18/3, pp. 297–312.

Laslett, P. (1991): A fresh map of life: The emergence of the third age, Harvard: University Press.

Llewellyn, J./Chaix-Viros, C. (2008): The Business of Ageing: Older Workers, Older Consumers: Big Implications for Companies. London, http://www.nomuraholdings.com/csr/news/data/news30.pdf (last access: 19.04.2013).

McLean, A. (2011): "Ethical frontiers of ICT and older users: cultural, pragmatic and ethical issues." In: Ethics and information technology 13/4, pp. 313-326.

Mantovani, E/De Hert, P. (2010): The EU and the e-inclusion of older persons. In: Mordini, E./de Hert, P. (eds.), Ageing and Invisibility, Amsterdam: IOS Press, pp. 1-50.

Masoro, E.J. (2006): "Are age-related diseases an integral part of aging?" In: Masoro, E.J./Austad, S.N. (eds.), Handbook of the biology of aging, London: Elsevier, pp. 43–62.

Ministerial Declaration of Riga (2006): EU Member States' Ministries responsible for media and communication, Ministerial Declaration, http://ec.europa.eu/information_society/activities/ict_psp/documents/declaration_riga.pdf (last access: 20.01.2015)

MIPAA/RIS (2002): United Nations. Economic and Social Council. Regional Implementation Strategy (RIS) for the UNECE region, Berlin: ECE/AC.23/2002/2/Rev.6.

Moody, H.R. (1994): "Four scenarios for an aging society." In: Hastings Center Report 24/5, pp. 32-35.

Mordini, E./de Hert, P. (2010): Ageing and Invisibility, Amsterdam: IOS Press.

Moulaert, T./Paris, M. (2013): "Social policy on ageing: The case of 'active ageing' as a theatrical metaphor." In: International Journal of Social Science Studies 1/2, pp. 113-123.

Natali, D./Vanhercke, B. (eds) (2013): Social developments in the European Union 2012 ETUI, Brussels: OSE Publications.

Organisation for Economic Co-operation and Development (OECD) (1998): Maintaining Prosperity in an Ageing Society, Paris: OECD.

Plomer, A. (2008): "The European Group on Ethics: Law, politics and the limits of moral integration in Europe." In: European Law Journal 14/6, pp. 839-859.

Robine, J.M./Mathers, C.D./Bone, M.R./Romieu, I. (eds.) (1993): Calculation of health expectancies; harmonization, consensus achieved and future perspectives/Calcul des espérances de vie en santé : harmonisation, acquis et perspectives, Paris: John Libbey Eurotext.

Rowe, J.W./Kahn, R.L. (1997): "Successful Aging." In: The Gerontologist 37/4, pp. 433-440.

Scruton, R. (2012): "Timely Death." In: Philosophical Papers. 41/3 pp. 421-434.

Searle, J. (1995): The Construction of Social Reality, New York: The Free Press.

SENIOR project (2008): Deliverable D1.1. Environmental scanning report, http://www.cssc.eu/public/REPORT%20for%20external%20on%20Environmental%20Scanning.pdf (last access: 12.01.2015)

Stahl, B.C. (2011): "IT for a better future: how to integrate ethics, politics and innovation." In: Journal of Information, Communication & Ethics in Society 9/3, pp. 140-156.

Tallacchini, M.C. (2009): "Governing by values. EU Ethics: soft tool, hard effects." In: Minerva 47, pp. 281-306.

Treaty on the Functioning of the European Union (TFEU) (2009): Consolidated Version of the Treaty on the Functioning of the European Union, 2008 O.J. C 115/47.

Trichet, J. C. (2010): "State of the Union: The Financial Crisis and the ECB's Response between 2007 and 2009." In: Journal of Common Market Studies 48/1, 7-19.

VALUE AGEING (2014): Incorporating European Fundamental Values in ICT for Ageing. Project funded by the European Commission-FP7 Marie Curie Industry-Academia Partnerships and Pathways Action, http://www.value-ageing.eu/ (last access: 12.01.2015)

VALUE AGEING (2013): Deliverable 7.1. WP Overview and Implementation Plan, http://www.value-ageing.eu/. (last access: 12.01.2015)

Walker, A. (2002): "A strategy for active ageing." In: International social security review 55/1, pp. 121-139.

Walker, A. (2009): "Commentary: The emergence and application of active aging in Europe." In: Journal of Aging and Social Policy 21/1, pp.75-93.

Winner, L. (1980): "Do Artifacts Have Politics?" In: Daedalus 109/1, pp.121-136.

World Health Organization (WHO) (2002): Active Ageing: A Policy Framework, Geneva: World Health Organization.

Ageing and Technology Decision-making: A Framework for Assessing Uncertainty

CIARA FITZGERALD AND FRÉDÉRIC ADAM

1. INTRODUCTION

We are living in an innovation age challenged with increasing complexity of decisions in all aspects of society. Such ICT innovations, advancing at an unprecedented rate, demand a sophisticated policy response to assess the impact of the rapid technological advances on society. We are keen to investigate the different types of uncertainty facing policy makers in the context of ICT innovation and the Ageing Society. This study provides a wide-ranging analysis of European policy relating to telecare and telemedicine. The purpose of our study is to explore policy measures related to telecare and home-based telemedicine in the European countries represented by a consortium of 14 European countries.[1] This case is of interest as European countries are challenged with an ageing population. Specifically, between 2000 and 2050, the proportion of the world's population over 60 years will double from about 11% to 22%. The absolute number of people aged 60 years and over is expected to increase from 605 million to 2 billion over the same period. The world will have more people who live to see their 80s or 90s than ever before. The number of people aged 80 years or older will have almost quadrupled to 395 million between 2000 and 2050. Indeed the probability of needing care increases with age. Less than 1% of those younger than 65 years need long-term care, while 30%

---

1 | PACITA is a four-year EU financed project under FP7 aimed at increasing the capacity and enhancing the institutional foundation for knowledge-based policy-making on issues involving science, technology and innovation, mainly based upon the diversity of practices in Parliamentary Technology Assessment (PTA).http://www.pacitaproject.eu/

of the women aged 80 years or over use long-term care services, on average across the OECD (OECD 2011). In light of these trends, recent innovative developments in technology have produced ICT devices to address the many challenges in supporting elderly people.

Information and communication technology (ICT)-based care technologies include real-time audio and visual contacts between patients and caregivers; embedded technologies such as smart homes, clothes and furniture to monitor patients inside and even outside their homes; electronic tagging of dementia patients and more biotechnological innovations, such as implants and devices for chronic disease monitoring. These technologies cover a wide range of innovations, from those already functioning to those that are prospective and theoretical. They provide health care and enable elderly people to maintain their autonomy and allow them to live independently for a longer period of time. These technologies are subsumed under the term telecare. However as well as the positive benefits, theorists are speculating on the social and legal risks of telecare, specifically regarding the issue of technology failure and the onus of responsibility, be it the users or the providers of the technology (Percival/Hanson 2006). It is worth noting the intention of this chapter is not to determine the ethics of telecare. Rather, this chapter supports the call by Yanga/Zhiyong Lan (2010) for the need for further study to facilitate our understanding of efficient policy making for how technology can address the challenges of the Ageing Society.

Given that telecare technologies is a ripe area of innovation which will have positive and potentially challenging societal implications, our research objective is to explore the different types of uncertainties for policy makers addressing how technology can address the challenges of the Ageing Society. We situate this study against the backdrop of 'responsible innovation', a growing scholarly appreciation that the advancements in ICT should be positioned within a societal context focused on the future consciousness of social well-being. Responsible Innovation is recognised to be a dynamic concept enacted at multiple levels and is forecasted to feature on the political agenda in the coming years. The term Responsible Innovation is defined as "taking care of the future through collective stewardship of science and innovation in the present" (Stilgoe/Owen/Macnaghten 2013: 3).

Our study contributes to this paradigm as it explores policy makers' response to telecare innovation and explores to what extent policy is considering the opportunities and challenges of telecare within the context of the aging society wellbeing. Within this agenda of 'Responsible Innovation', our chapter will outline how decision support can facilitate better technology assessment processes, which are needed to manage innovations in ICT to address the Ageing Society. The rest of the chapter is organised as follows. The next section presents the literature review focusing on how technology can address the challenges of the Ageing Society and decision support. Following this, we outline the methodology, then the findings, followed by a discussion of the findings before concluding with outlining implications for policy makers.

2. LITERATURE REVIEW

ICT Policy Making

ICT policy making is a much studied area for scholars, as countries grapple with new innovative technologies and question their impact. Governments are challenged to scientifically assess societal, ethical, legal and economic aspects of technology. However, Delvenne et al. (2011) argues uncertainty is no longer contained within modern structures of policy making. Specifically, they argue the current challenge for policy makers is to accommodate the uncertainty and dynamics of patterns to offer the decision- making process "a context-determined and temporally limited orientation for action that makes learning through experience possible" (Delvenne et al. 2011: 18). Furthermore, no discussion on such complex decision making is meaningful without a discussion of the limitations which apply to human decision making, as described by Simon under the term of bounded rationality (March/Simon 1958).

Scholars argue under conditions of bounded-rationality, decision makers seemingly "do what they can" or in some cases, "make-do". To further complicate matters, when contemplating boundaries in ICT policy decision making, traditional boundaries are not imposed, but constructed, bargained, negotiated and appropriated by stakeholders (Delvenne et al. 2011). We speculate such boundary-less domains can result in ambiguity of decision making within a fluid environment. Policy makers need to urgently respond to demands of citizens to engage more pro-actively in policy decisions that heavily concern particular stakeholder groups and citizens (Wimmer et al. 2012). We question if policy makers as decision makers in such fluid environments are indeed 'muddling through' according to Lindblom (1959). If so, we propose a decision support lens which can make a potentially powerful contribution and will provide recommendations of real pragmatic value for policy makers. Our study will address the need for more clearly defined and systematic theoretical and empirical studies to facilitate our understanding of efficient policy decisions (Yanga/Zhiyong Lan 2010)

Decision Support

We propose a decision support lens will help improve the quality of decision making for policy makers addressing telecare technology. Decision support systems can be effective in multiple environments for a range of uncertainties. Following Alter, one should explore how decision support can be provided when considering the feasibility of Decision Support Systems (DSS). Alter (1992, 2004) has repeatedly pointed out that the development of DSS was secondary to the objective of improving the quality of decision making, calling for a focus on decision support rather than decision support systems. To provide decision support, one should concentrate on developing an overall system of decision making which is based on evidence and supported by expert advice. The use of DSS in a political context can be problematic as studies found that the inherent rationality of the DSS was in conflict with how participants usually make decisions as well as with the political process (Andersson/Gronlund/Astrom 2012). For an in depth analysis of DSS and a framework for the classification of DSS usage across organizations see Adam et al (1998). We are keen to further explore the feasibility of DSS in a policy making context for how technology can address the challenges of the Ageing Society. We argue policy decision making for how technology can address the challenges of the Ageing Society addresses a number of categories of uncertainty (Earl/Hopwood 1980). Specifically policy making for ICT innovation considers the following:

- Uncertainty about the mechanics of technologies – the what question
- Uncertainty about their impact – the who and how questions
- Uncertainty about societal preferences – the why questions

Earl and Hopwood (1980) have theorised on the nature of uncertainty and, leveraging Thompson and Tuden (1959), they have distinguished uncertainty about the cause and effect relationship versus uncertainty which relates to the preferences of the stakeholders. Silver (1991) also proposes a reflection on the difference between guidance underpinned by information and guidance aimed at prescribing choices, which he respectively labels informative decisional guidance and suggestive decisional guidance. In general, DSS applications must rely on the existence of clear modelling and reasoning to underpin the optimisation algorithms that are being applied. DSS are applicable to all types of stakeholders, both managerial and non managerial and are used to support fully structured problems where it is possible to specify algorithms or decision rules that allow the problem to be found, alternative solutions to be designed and the best solution to be selected (Adam et al. 1998). The key issue is therefore whether societal decisions in the area of ICT lend themselves to the development of what Earl and Hopwood (1980) term answer machines and what happens when the level of uncertainty and ambiguity involved means that the provision of the answer machine can potentially compromise the ability of policy makers to make the right choices. Earl and Hopwood (1980) have warned against trying to hide the true complexity of societal problems (focusing on developing DSS), rather than embracing it (focusing on improving the quality of decisions). Where assumptions are made about the future, or where consensus has not yet arisen in an organisation or society, decision support should not provide artificially complete ready-made answers and should, instead, promote judgment and dialogue amongst stakeholders. Although the concept of decisional guidance may appear intangible, Earl and Hopwood's (1980) recommendations provide tangible avenues for analysing decisional guidance in terms of its fit with the problems facing policy makers. We propose therefore that certain societal problems with a given technology can lead to suggestive guidance, whereas others cannot and should not, given the state of development of policy-makers' understanding or the absence of a clear societal consensus. This has clear implications for the type of dialogue which must take place in society in relation to different types of innovations. This chapter explores policy making in the area of telecare technology as an example of ICT innovation and considers how DSS can support policy making, be it as suggestive guidance or otherwise.

3. METHODOLOGY

The methodology utilized in this study is an in-depth case study approach of European response to telecare policy. The case-study method has been widely recommended for study areas that are not yet well understood and lack formal theories (Yin 2009). It is particularly relevant for our study as there are very few studies on the actual use of DSS in a political context (Andersson/ Gronlund/Astrom 2012). It has also gained particular popularity in the public policy literature because of 'the depth and richness' the result can provide for enlightened public policies (Silverman 2013). Since ICT innovation and decision support is a comparatively new and underexplored policy issue, a case study approach can provide rich context-dependent knowledge to assist policy-making. The fourteen countries in the study were Austria, Belgium, Bulgaria, The Czech Republic, Denmark, Germany, Hungary, Ireland, Lithuania, Norway, The Netherlands, Portugal, Spain and Switzerland. The selection of countries signifies the participating partners in a FP7 funded project called PACITA and represents Europe as our case study. The categories used to capture consistent data on each country, were as follows: Definitions, National Demographic Trends, National policies, Policy Enablers, Policy Enactors, Actor Involvement Incentives, Service Providers, Technologies in use, and a Risk Analysis. The choice of categories for inclusion were grounded in relation to its practical purposes following recommendations by studies analyzing and comparing how technology can address the challenges of the Ageing Society (Ishmatova/Thi Thanh Hai 2013).

Specifically, we explored the definitions used in policy documents in the 14 European countries in the study. Then, we assessed demographic conditions. Next, we examined specific national policies. Following this, we investigated the key actors involved. We categorized them as policy enablers and policy enactors. On consultation with national experts, there appears to be many risks, but these are not recognised in policy discussions thus far. Extensive desk research was conducted for each country. To complement this, policy experts were contacted for additional information that was not easily accessible via secondary sources. The richness of information differed from each country; however, this is not problematic as it reflects their differing levels of policy sophistication in the area of telecare. Therefore, there are some apparent nuances in the approaches but this adds to the complexity of the findings. For the purpose of this study, relevant legal and policy documents, government publications and scholarly literature were examined, documenting developments up until September 2013. Documentary search and analysis were complemented by a series of semi-structured in-depth interviews. Finally, we held a workshop with key experts to validate our results, which led to further relevant analysis as outlined in the following section.

It is worth noting that the purpose of the chapter is not to compare and contrast national approaches to telecare but indeed to explore how the challenge of ICT and the ageing society can be addressed using DSS. Therefore, we present the findings at an aggregate level rather than individual and offer examples to best represent the nuances in the richness of the data. We will present the findings in the form of the key questions following Earl and Hopwood (1980) framework that can be addressed using a DSS.

4. CASE STUDY

The primary function of telecare technology is to address the challenge of an ageing population. Our study revealed there are a number of common societal challenges that emerged from our analysis. Specifically, the countries in our study are challenged with a significant increase in life expectancy, an increase in senior citizens suffering from dementia and other age related illnesses, rising cost of care of senior citizens and an increased demand for independent living solutions. In light of such challenges, policy makers are exploring emerging technological advances in telecare and telemedicine to offer solutions. However, the technologies are not an exclusive panacea for the aforementioned challenges and indeed pose responsibilities for policy makers with regard to uncertainty over cause and effect of the innovative technology. The findings provide a basis for commentary and serve to promote awareness of the policy status in telecare in Europe, as represented by the European countries in our study. We present the findings following the categories of uncertainty, presented in the literature review. Specifically, we discuss uncertainty about the mechanics of technology, uncertainty about their impact and uncertainty about societal preferences.

Uncertainty about the Mechanics of Technologies – the What Question

Our findings reveal common interpretations of telecare and telemedicine are used in national documents; however, they are used interchangeably in many contexts. There are a number of interrelated concepts such as Ambient Assisted Living, eHealth, Assistive Technology, ICT in Health, Welfare Technology and Telehealth. Such level of uncertainty is not conducive to effective policy making. European policy makers need to determine a definitive understanding of what telecare technology is. Furthermore, there are differing levels of sophistication regarding the formulation and implementation of telecare policy. Firstly, there is difference in timeline in the adoption of telecare policy. For example, 1993 was the earliest policy initiative in Austria whereby a tax funded long-term care system that is independent from income was introduced. Since then, all countries in our study have documents referencing telecare, or equivalent but to varying levels of comprehensiveness. For example, the policies are at various levels of a continuum concerning frameworks for security and strategies for encouraging adoption. We argue some countries are proactive in seeking opportunities for encouraging the development of telecare, whilst others are reactive and seek only to fulfil the minimum requirements of regulation. The following are areas of uncertainty about the mechanics of technologies that arose from our analysis.

- What is telecare technology?
- What are the best policies to encourage the development of telecare?
- What are the regulatory requirements?

Uncertainty about their Impact – the Who and How Questions

Our findings reveal evidence of fragmented, uncoordinated decision-making and implementation in the telecare domain with no central responsibility for policy making in all countries in our study. Specifically we addressed Policy Enablers, a categorization of actors responsible for the formulation of telecare policy and Policy Enactors, a categorization of actors responsible for the implementation of telecare policy.

In the category of Policy Enablers, our findings highlighted a surge in the number of government departments getting involved in the telecare domain. For example, the Ministry of Health, the Ministry of Social Affairs, and the Ministry of Industry and Employment were all recognised to play a role in some aspect of telecare, in various countries. Due to the various groups involved, we can deem telecare policy to be a complex policy making subject. To add to the complexity, there is also an additional dimension of regional versus national policy, for example in the case of Belgium. We argue there is no single group taking responsibility for the formulation of telecare policy and this can be deemed a weakness.

In the category of Policy Enactors, our findings reveal there is a mix of non-profit, voluntary, and non-governmental agencies involved. For example, there are age action groups, charitable organisations, age support groups. Their roles are varied and include raising awareness, and dissemination of research. Similarly, we argue there is a sense of unaligned discourse in the implementation of policies relating to telecare. When researching telecare service providers, we found private firms to be dominant in both service and product offerings. It is also a space that lends itself to innovation for new companies. There are a growing number of startups in the telecare field and it is particularly common area found among spinouts from universities. We propose this may be due to the large funding programmes such as FP7. An example of such a university spin out is Technology in Healthcare, spun out from Bangor University in 1998 in Wales, which was responsible for TED, The Electronic Doctor, and a number of other smart sensor products.

Another, particularly interesting case is the Hagen Committee in Norway. This is a national program for municipal innovation in care where 1% of care services budget is allocated to Innovation in the form of a Private public partnership. The implication of such an initiative encourages dedicated innovation in a particular phenomenon and has a largely positive contribution to the innovative capabilities of SMEs and start-ups seeking to develop their technology for the ageing society.

However, other countries have yet to formalize initiatives to encourage innovation between private and public institutions. This is an area that could be further explored. The following are areas of uncertainty about the impact that arose from our analysis.

- Who should be responsible for the formulation of telecare policy?
- Who should be responsible for the implementation of telecare policy?
- How can public- private partnerships encourage advancement of telecare technology?

Uncertainty about Societal Preferences – the Why Questions

Surprisingly, the societal preferences of telecare were largely absent in national telecare policy documentation. At the workshop with experts we were engaged in an interesting discussion on the risks associated with telecare. Here, several types of risks were identified, including Privacy Risk, Social Risk, Technology Risk, Legal Risk and Financial Risks. The most common privacy risks were concerned with legal rights and ethical considerations not being fully addressed by policy makers. The social risk of isolation was considered and the question of forced or voluntary participation was raised. Such an issue has serious implications on the societal fabric of a country and should be considered at a national level.

The technology risk of how to secure data storage and transmission of sensitive health data were identified. Such an issue is not exclusive to telecare as indeed Big Data is a term today that impacts all areas of technology and society including social media data, location data generated by smart phones and other roaming devices, public information available online and data from sensors embedded in cars, buildings and other objects.

Also the polarized dilemma of technology driven innovation versus user need innovation was questioned and the ramifications of this debate for policy making. Whether technology should be driven by societal need or technological capability is an issue policy makers need to address.

Legal risks were also articulated; specifically the medical responsibility was questioned in the technology versus practitioner onus of responsibility debate in the time of malpractice. The legal risk of the lack of legislation and regulation in this space was also recognised. Finally, the financial risk was discussed as to the question of who is responsible for the costs of the telecare technology. The following are areas of uncertainty about the societal preferences which arose from our analysis.

- Why telecare risks are not being discussed at policy level?

5. IMPLICATIONS OF RESEARCH

As evident from our multiple country study, decision making regarding telecare policy is a fragmented, challenged process, with differing levels of sophistication. Our interest lies in decision support as a provision to holistic policy making in telecare to address the grand challenge of ageing. We argue policy makers, when challenged with policy making in how technology can address the challenges of the Ageing Society are suffering from a crisis of legitimacy as evident from the different types of uncertainty. As articulated by Kovisto et al. (2009: 1164), innovation processes have shifted from 'the positivist and rationalist technology-focused approaches towards the recognition of broader concerns that encompass the entire innovation system, including its economic, social and economic perspectives'.

As there are nuances among countries in their policy efforts, we argue decision support can frame the uncertainty over preferences and reduce uncertainty over cause and effect. Carter and Bélanger (2005) argue government agencies must understand the factors that influence citizen adoption of innovation. Their findings indicate that perceived ease of use, compatibility and trustworthiness are significant predictors of citizens' intention to adopt technology.

Similarly, other studies highlight trust as the key success factor in technology acceptance of multi-criteria decision-support systems in the case of high impact decisions (Maida et al. 2013). We advance this argument and argue decision support can help promote judgment and dialogue with citizens thus providing rich material. In support of our argument, Rose and Grant (2010) argue that involvement from all stakeholders, including citizens of various ICT means and capabilities is a requisite for successful implementation.

We propose a research agenda to explore further decision support mechanisms to support how technology can address the challenges of the Ageing Society. We propose a number of implications are to be considered as evident from the findings of our case study of European telecare policy. As shown in our case study, none of the 14 countries have a dedicated policy for telecare. Whilst all recognize their national demographic trends demand a telecare response, there are a wide variety of responses in how each country in our study are engaging with telecare policy. A decision support response can frame the uncertainty and present suggestive decisional guidance following Earl and Hopwood (1980) recommendation, thus instilling trust and legitimacy in the policy making process.

Our findings recognize the challenge of applying a DSS to a decision in complicated and contested matters such as the use of technology in the aging society yet we support a DSS offering of suggestive guidance with the following two caveats. Firstly, policy- makers need to understand the clear societal consensus, and secondly, where assumptions are made about the future, or where consensus has not yet arisen in society, decision support should not provide artificially complete ready-made answers and should, instead, promote judgment and dialogue amongst stakeholders (Earl/Hopwood 1980).

Furthermore, decision support can increase integrity and honesty in policy decisions, two vital components to the success of transformation of policy making for technology innovation. Public sector values are the foundation from which the idea of genuine transformation ultimately derives (Bannister/Connolly 2014). The second implication of a decision support framework will promote a sense of action thus ensuring a sense of positivity about how technology can address the challenges of the Ageing Society policy decision making. The final implication will translate a respect for the citizen. Decision support for policy making in ICT innovation will support moving beyond a utilitarian and unidirectional approach to technology, thus fostering engagement through institutionalization of citizen engagement and debate on contentious issues in ICT through increased transparency in the outcomes of decisions (Evans/Campos 2013).

6. Concluding Comments

The chapter will be of interest to Aging Society scholars, policy makers, and society in general as we explore a framework for assessing uncertainty in ageing and technology decision-making. Specifically, our study provides a picture of the uncertainty in policy making relating to how technology can address the challenges of the Ageing Society, and lends itself to further study of how decision support can frame uncertainty. We have considered 14 countries and their policy approaches to telecare. We suggest that contemplating decision support will frame uncertainty and deliver a number of implications, including legitimacy of policy, infer a sense of action and deliver a respect for the citizen. Our framework supports the feasibility and desirability of shaping and steering decision support in how policy-making can influence technology to address the challenges of the Ageing Society.

We believe that the framework we developed in this study is a significant development of previous work in the DSS area. In particular, our framework suggests that focusing on the role of DSS in ICT policy making is, potentially at least, a much richer vein for DSS research than focusing on specific types of problems or individual decision support systems.

Our study also contributes to the growing area of interest of Responsible Research and Innovation. Citizens, policy makers and funding agencies are seeking an assurance of responsibility when introducing innovation to society. DSS can offer a validity and transparency to decisions in contentious areas of innovation.

Further research should attempt to verify that, as policy makers use DSS applications to support their ICT and Ageing Society problem solving, their knowledge of these problems increase to such an extent that these problems become increasingly specifiable for the national context. Longitudinal studies of DSS usage could give a more accurate indication of the extent to which ICT and the Ageing Society challenges can be addressed from the implementation and refinement of decision support applications.

REFERENCES

Adam, F./Fahy, M./Murphy, C. (1998): "A framework for the classification of DSS usage across organizations." In: Decision Support Systems 22/1, pp. 1-13.

Alter, S. (1992): "Why persist with DSS when the real issue is improving decision making?" In: Jelassi, T., Klein, M.R., Mayon-White, W.M.(1992) Decision Support Systems: Experiences and Expectations. , North Holland. IFIP

Alter, S. (2004): "A work system view of DSS in its 4th decade." In: Decision Support Systems 38/3, pp. 319-327.

Andersson, A./Gronlund, A./Astrom, J. (2012): "You can't make this a science – Analysing decision support systems in political contexts." In: Government Information Quarterly 29/4, pp. 543-552.

Bannister, F./Connolly, R. (2014): "ICT, public values and transformative government: A framework and programme for research." In: Government Information Quarterly 31/1, pp. 119-128.

Carter, L./Bélanger, F. (2005): "The utilization of E-Government services: Citizen trust, innovation and acceptance factors." In: Information Systems Journal 15/1, pp. 5–25.

Delvenne, P./Fallon, C./Brunet, S. (2011): "Parliamentary technology assessment institutions as indications of reflexive modernization." In: Technology in Society 33/1–2, pp. 36-43.

Earl, M.J./Hopwood, A.G. (1980): "From management information to information management." In: Lucas, H./ Land, F./ Lincoln, T./ and Supper, K. (eds.), The Information Systems Environment, North-Holland, IFIP, pp. 133-143.

Evans, A.M./Campos, A. (2013): "Open Government Initiatives: Challenges of Citizen Participation." In: Journal of Policy Analysis and Management 32, pp. 172–185.

Ishmatova, D./Thi Thanh Hai, N. (2013): "Towards a framework for analysing and comparing ICT policies for Aging Society Policies: A First Approximation." In: ICEGOV'13, October 22-25 2013, Seoul, Republic of Korea.

Koivisto, R./Wessberg N./Eerola A./Ahlqvist, T./Kivisaari, S./Myllyoja, J./ Halonen, M. (2009): "Integrating future-oriented technology analysis and risk assessment methodologies." In: Technological Forecasting and Social Change 76, pp. 1163-1176.

Lindblom, C. (1959): "The science of 'muddling through." In: Public Administration Review, 19(2) pp. 779-88.

Maida, M./Maier, K./Obwegeser, N./Stix, V. (2013): "Success of multi criteria decision support systems: The relevance of Trust." In: 46th Hawaii International Conference on System Systems, pp.1530 – 1605.

March, J./Simon, H. (1958): Organisations. New York. J Wiley

OECD (2011): Live Longer, Work Longer. OECD. Paris.

Percival, J./Hanson, J. (2006): "Big brother or brave new world? Telecare and its implications for older people's independence and social inclusion." In: Critical Social Policy 26/4, pp. 888-909.

Rose, W. R./ Grant, G .G., (2010): "Critical Issues pertaining to the planning& implementation of e-government initiatives." In: Government Information Quarterly 27/1, pp. 26-33.

Silver, M.S. (1991): "Decisional guidance for computer-based decision support." In: MIS Quarterly 15/1, pp. 105-122.

Silverman, D. (2013): Doing qualitative research: A practical handbook. London. Sage.

Stilgoe, J./Owen, R./Macnaghten, P. (2013): "Developing a framework for responsible innovation." In: Research Policy 42/9, pp. 1568-1580.

Thompson, J. D./Tuden, A. (1967): Strategies, structures, and processes of organizational decision. New York. Bobbs-Merrill.

Wimmer, M./Scherer, S./Moss, S./Bicking, M. (2012): "Method and Tools to Support Stakeholder Engagement in Policy Development: The OCOPOMO Project." In: International Journal of Electronic Government Research, 8/3, pp. 98-115.

Yanga, L./ Zhiyong Lan, G. (2010): "Internet's impact on expert–citizen interactions in public policymaking – A meta analysis." In: Government Information Quarterly 27/4, pp. 431-441.

Yin, R.K. (2009): Case study research: Design and methods. London. Sage.

Aging and Technology: What is the Take Home Message for Newspapers Readers?

GREGOR WOLBRING AND BOUSHRA ABDULLAH

1. INTRODUCTION

Worldwide the number of older persons will exceed the number of young persons in 2050 (United Nations Department of Economic and Social Affairs Population Division 2012; United Nations Department of Economic and Social Affairs Population Division 2001), a development seen to impact many aspects of life including the equity and solidarity within and between generations (United Nations Department of Economic and Social Affairs Population Division 2001). Various documents suggest actions that have to be taken for ensuring a positive environment for aging (Second World Assembly on Ageing 2002; World Health Organization 2002): ageism, the negative social treatment experienced because of one's age, is discussed extensively as a threat to aging well (Angus/Reeve 2006).

Technologies have long been identified as having an impact on aging and the elderly: as early as 2002, the Political Declaration and Madrid International Plan of Action on Ageing states in its introduction that one should focus on "realizing the potential of technology to focus on, inter alia, the individual, social and health implications of ageing, in particular, in developing countries" (Second World Assembly on Ageing 2002: article 12). It highlights the existence of technologies that can be used to promote independence, to bring people together, to reduce marginalization, loneliness and segregation between the ages, and to generate positive socioeconomic changes. It questions the lack of access to these technologies by many old people and asks as an action item for "measures that enable older persons to have access to, take part in and adjust to technological changes should therefore be taken" (Second World Assembly on Ageing 2002: article 38).

The impact of technologies on aging and the elderly is also reflected in the amount of academic work generated on the topic[1]. Academic inquiries into technologies are, among other issues, linked to innovation and consumerism within a framework of aging (Peine et al. 2014), ageism being the negative treatment of people based on their old age (Cutler 2005); intergenerational equity (Binstock 1992); intergenerational justice (Gusmano/Allin 2014); intergenerational cooperation (Heath 2013); active aging; design of next generation portable supporting devices (Chan et al. 2014) and aging in place (Peek et al. 2014). As to specific technologies, two recent additions to the technology and aging discourse are the fields of social robotics – with application ideas ranging from health care to being companions for the elderly (Flandorfer 2012; Heerink et al. 2010; Wolbring/Yumakulov 2014; Yumakulov et al. 2012) – and home care technology, where the envisioned applications range from monitoring the elderly to the ability of the elderly to self-care (Berridge et al. 2014; Wolbring/Lashewicz 2014).

Although academics are intensively engaged with the topic of aging and technology, how does the public gain access to this knowledge – given that most academic literature is not physically accessible to the public? Even if such information is available, how can the general public know what keyword to use in an internet search engine to find the data if they are not exposed to the keyword (Wolbring et al. 2012)? Distribution of knowledge through printed media is seen as an essential part of enabling social participation (Nord 1988) and sustaining political freedom and a stable social order (Weinstock Netanel 1996). Media often set discussion agendas for societies and create the boundaries within which debate takes place (Tynedal/Wolbring 2013; Wallack 1990). Various aspects of newspaper coverage of aging have been studied, such as the portrayal of aging and the elderly (Buchholz/Bynum 1982; Matcha/Sessing-Matcha 2007), elder abuse (Mastin et al. 2007), successful aging (Rozanova 2010), nursing homes (Miller et al. 2012), constructing aging (Fealy et al. 2012), Alzheimer (Kang et al. 2010), attitudes towards aging (Hubbard et al. 2009), the elderly in the workforce (Powell 2013), construction of the subject of care (Weicht 2013), aging and healthcare spending (Gusmano/Allin 2014), and the visibility of the 2002 *World Health Organization report Active Ageing: a policy framework* and

---

1 | Google scholar, a database covering academic work, generated 1,950,000 hits with the keyword combination of 'elderly' and 'technology', 2,070,000 hits with 'aging' and 'technology', and 652,000 hits with 'ageing' and 'technology'.

the *2010 Toronto Charter for Physical Activity: A Global Call for Action* (Abdullah/Wolbring 2013).

As to the coverage of technologies in relation to aging and the elderly, one article assessed the quality of newspaper medical advice columns for elderly readers (Molnar et al. 1999). However, no analysis of how technologies as a whole are covered by newspapers in relation to the elderly and aging could be found. We therefore investigated the discourse around technology and aging and the elderly in two Canadian newspapers; one with national reach (*The Globe and Mail*) and one with local reach from the Canadian province of Alberta (*Calgary Herald*) from 1980-2013. We report qualitative and quantitative data of a study that asked two questions: 1) Which technologies were mentioned in the newspapers in regard to aging and the elderly and 2) How were technologies related to aging or the elderly covered (utility, problems, social groups linked to them, ethical issues and technology mentioned in relation to medical or social issue faced by the elderly…)?

We discuss the findings through the lens of two Canadian policy reports on aging, one being the 2009 Canadian Senate report on aging *Canada's Aging Population Seizing* the *Opportunity* (hereafter cited as the Canadian report) (Canada. Parliament. Senate. Special Committee on Aging et al. 2009) and the other being the 2013 report from the Province of Alberta *Let's talk about aging: aging well in Alberta* (hereafter cited as the Alberta report) (Chief Medical Officer of Health Alberta 2013) and through the lens of an older document that proposed various actions the 2002 Madrid International Plan of Action on Ageing (hereafter cited as the Madrid action plan) (Second World Assembly on Ageing 2002), all of which mention media as an essential part of improving the situation of aging and the elderly. We chose the Madrid action plan as it was older and the state of technologies was different than when the Alberta and the Canadian report were written.

2. METHOD

Data Source and Sampling

To obtain quantitative data on the frequency of the term "technology" together with the terms "elderly", "aging" or "seniors", the following newspapers were searched on May 25, 2014 for the years 1980-2014 (table 1): *The Globe and Mail* (Canada, national reach), *Calgary Herald* (Canada, local reach) and the *Canadian Newsstand Complete*, a collection of 300 Canadian newspapers. We accessed *The Globe and Mail*, the *Calgary Herald* (Canada) and the *Canadian Newsstand Complete* through Proquest databases, which in turn we accessed through the University of Calgary.

To obtain qualitative and quantitative data on the discourse around aging and technology, we investigated *The Globe and Mail* and the *Calgary Herald* from 1980-2014. Using the keyword combination "elderly" and "technology", we obtained n=935 *The Globe and Mail* articles and n=365 *Calgary Herald* articles. Using the keyword combination "aging" and "technology", we obtained n=2225 *The Globe and Mail* articles and n=1167 *Calgary Herald* articles. We used the spelling "aging", as that was the dominant spelling over "ageing" in the Canadian newspapers. Articles identified were downloaded as PDF and imported into Atlas-TI (May 25, 2014), a qualitative data analysis and research software.

Data Analysis

The research questions we investigated were the following: 1) Which technologies were mentioned in the newspapers in regard to aging and the elderly and 2) How were technologies related to aging or the elderly covered (utility, problems, social groups linked to them, ethical issues, etc.). We employed deductive and inductive coding strategies. As for deductive strategies, we used the Word Cruncher function of Atlas-TI to generate a hitcount of all the words evident in the newspapers, and we analyzed the list of words for relevance to our research questions. We also employed an inductive and iterative coding strategy, in which articles were read and themes relevant to the research questions were identified as reading progressed. When we started reading the articles, we realized that most articles did not really cover technologies, as they related to aging or the elderly, but had by chance both terms in the same article. Therefore, we employed the co-occurrences function of Atlas-TI where the software searched for all the incidents in which both terms appeared in the vicinity (same paragraph) of each other. We searched all of *The Globe and Mail* and *Calgary Herald* articles for the mentioning of aging and technology (n=222, n=126 relevant) or elderly and technology (n=56, n=53 relevant) in the same paragraph in order to obtain more relevant content[2]. All paragraphs

---

2 | We searched for paragraphs that co-contained the terms abil* and aging (n=57); abil* and elder* (n=17); abil* and technol* (n=43); cost and tech-

identified by the software were read by both authors and coded with the research questions in mind to increase reliability, and differences were resolved during our discussions.

Limitation

We did perform an in-depth content analysis of only two English language Canadian newspapers. As such, our findings are not generalizable for Canada or other countries: our data cannot be used to judge other media types either, as we focused on newspapers. We also used the term technology to identify articles and not terms of individual technologies. However, we content that our data can be used to guide future research in this area.

3. RESULTS AND DISCUSSION

General Coverage of Technology in Relation to Aging and the Elderly

It is well known that "newspapers are influenced by their environment, including ownership, funding, need for circulation, advertisement revenue and the readers preference for reading like-minded news." (Cheung/Wolbring 2015:82) However, the results we found with regard to the coverage of technologies in relation to aging and the elderly in Canadian newspapers are puzzling. We are living in a technological world and aging has been recognized for some time as an issue of interest to the Canadian public. Indeed this is reflected in the amount of Canadian newspaper articles that contain the words aging, elderly, seniors, elders and technology (table 1). The combinations of elderly and technology or aging and technology generate a substantial amount of articles, thus suggesting that technology is seen as an important issue as relates to aging and the elderly (table 1).

---

nol* (n=31); cost and elder* (n=23); cost and aging (n=13); technol* and policy (n=20); technol* and ethic (n=31); technol* and needs (n=151); technol* and healthcare (n=35); technol* and equity (n=3); technol* and accessibility (n=66); technol* and equality (n=3); technol* and seniors (n=7); technol* and problem (n=59) and technol* and women (n=24).

Table 1: Frequency of Co-occurrence of Technology with the Terms Aging, Elderly, Elders and Seniors

Keyword	The Globe and Mail	Calgary Herald	Canadian newsstand complete
Technology	127793	51702	1156522
Elderly	18714	9655	278573
+technology	935	365	9646
Aging	21364	12355	293338
+Technology	2225	1167	25543
Seniors	167644	77340	2191067
+Technology	15647	5451	122146
Elders	11710	6659	162349
+technology	562	279	6198

However, upon reading the articles it became evident that very few articles engage with technology as it relates to the elderly or aging. When we searched *The Globe and Mail* and the *Calgary Herald* articles for the co-occurrence of technology and aging/ageing in the same paragraph, we found only 126 relevant articles, and we found only 53 relevant articles that covered technology and the elderly in the same paragraph. Our findings suggest that the reader does not take home the message of the pervasiveness of the impact of technologies on aging and the elderly. This underwhelming coverage does not reflect, for example, the coverage of technology related to aging and the elderly in the academic literature. The three policy documents we use as a lens for our analysis also cover technologies extensively. The 2002 Madrid action plan mentions the term tech* n=60 throughout, the 2009 Canadian report contains the term tech* n=47 and has a whole chapter on technology (chapter 9) and the 2013 Alberta report mentions tech* n=27 times and has a section called "Technologies that support older adults and their caregivers" (17-18) and a box called "technical support" (18). Interestingly, the 2009 Canadian report was only mentioned in four of the n=300 newspapers of the *Canadian newsstand complete* database, one time each, with none covering the technology aspect of the report. The 2013 Alberta report has so far not once been mentioned by the 300 newspapers represented in the *Canadian newsstand complete* database, which includes the main Alberta newspapers, as has not the Madrid action plan.

As to the general application flavor of technology (medical or social narrative), the 2009 Canadian report focuses on technology and aging within the medical narrative and not on the use the elderly can make of all kinds of technologies outside of medical health narratives, while it singles out the utility of technology in "providing new opportunities to deliver care" (8). The 2013 Alberta report links technology applications to medical and social well-being as does the 2002 Madrid action plan.

Coverage of Specific Technologies in Relation to Aging and the Elderly

As to the coverage of specific technologies or applications, the relevant newspaper articles mention only twelve specific technologies or applications more than once.[3] None of these technologies were covered before 1990. With the exception of medical technology as a term, the coverage of these technologies was so low that no trends can be obtained. As for the term "medical technology", the coverage increased from the 1990-1999 to 2000-2009 decades, but the 2010-2014 number makes it less clear whether the increase in coverage will continue.

The lack of coverage of specific technologies and applications is also reflected in the 2009 Canadian report, which only mentions tele-health and electronic health records (which has its own sub chapter, 9.3.). As to these two technologies singled out by the Canadian report, telehealth or telemedicine was only once mentioned in *The Globe and Mail* in 2011, once in the 2002 Madrid action plan and not at all in the *Calgary Herald* and the Alberta report. Electronic health records were only mentioned once each in *The Globe and Mail*, the *Calgary Herald* and the Alberta report and not at all in the 2002 Madrid action plan. That the 2002 Madrid action plan does not mention telehealth might be understandable given its date, but it is surprising that the 2013 Alberta report does not thematize telehealth/telemedicine more especially in relation to home-care technologies, given that Alberta Health a government agency

---

3 | Robots/ robotic devices, n=5; assistive drivers or vehicles, n=8; for home/ living purposes, n=16; medical technologies, n=56; assistive technologies in public space, n=5; for anti- aging, n=10; for longevity/ living longer, n=6 technologies that assist caregivers/ nursing homes n=6 ; technology for physical activity, n=2; environmentally friendly technologies, n=2; genes and genetics and technology, n=2 and technologies specific to the aging process n=2.

had a unit on telehealth. However, in contrast to the Canadian report, the 2013 Alberta report mentions various other technology applications throughout the report such as monitoring technologies, home automation devices, medication reminders through phone or fax, sensors for detecting falls, home automation and GPS locators and internet use as social media to battle social isolation. The 2002 Madrid action plan mentions information and communication technology four times, farming technologies for aging farmers once, assistive technology linked to rehabilitation care twice, the term medical technology once, computer technology once and telemedicine once. However, academic literature shows that there are many more concrete technology applications applied to the elderly and aging. The readers of either the policy documents or the newspapers do not get an idea of how pervasive technology applications really are. Social robotics, which is an emerging field, is not at all on the radar screen yet, although the elderly are seen as a target group of consumers of both social robots (for assistance and companionship), and healthcare robots.

Themes Evident and Missing in the Coverage of Technology in Relation to Aging and the Elderly

Only one theme was covered in more than n=20 articles namely technologies and cost (n=44). Impact of technologies on the healthcare system was present (n=18) times. The following themes were only visible in less than ten articles such as, demand for technology (n=9), technology for increasing independence (n=8); adapting to technology (n=8) and n=6 articles mentioned the term problem. Many themes mentioned in the academic and grey literature around aging such as how technologies play themselves out around the aging of people from socially disadvantaged groups; ethics of technology and ageism linked to technologies (three themes whose lack of coverage we discuss further down) were hardly (three or less articles) or not all mentioned.

1 Theme present: Cost

As for code co-occurrence for technology and (aging and ageing) in the newspapers n=39 articles covered cost aspects[4]. As for the code co-occurrence results for Technology and elderly in the newspapers n= 5 were covering cost

---

4 | Before 1980 n= 0; 1980- 1989 n= 2; 1990- 1999 n= 19; 2000- 2009 n= 13; 2010-2014 n= 5.

aspects[5]. As to the role of technology in relation to cost and the elderly, some articles focus on technologies as ways to save money and some articles focused on technology as a cost driver with roughly even weight. Interestingly the theme of cost was mostly present in the newspaper articles published before 2002. As to the 2009 Canadian report cost is a main theme but only twice is cost linked to technology. The same is true for the 2013 Alberta report which covers cost in various areas but in regards to technology only twice; once with the focus of technology saving cost and once with the focus on increasing cost. The 2002 Madrid action mentions cost only in general and not once related to technology. Our finding suggest that although cost is a big theme it's much less discussed in relation to technology and if technology is covered the message seems to be evenly split between technology increasing and decreasing cost. We argue that the reader does not really get a good picture around the impact of technology on the topic of cost and the elderly.

In the following we give some more qualitative details on how cost was covered in the newspapers and the two Canadian policy documents. As to the coverage of the term "cost" mentioned in the vicinity of "elder*", "aging" or "seniors", one finds topics such as the cost of the elderly (Armstrong 2002), the increased cost of the pension plan (Speirs 1983), the cost of access to healthcare, overuse of drugs, and one article assumes a negative impact if a cure for cancer would appear. Related to articles covering the terms "technol*" and "aging"/the "elder*" and "cost", articles mentioned the focus of a venture capital fund on low tech products for elderly care (*The Globe and Mail* 1989), the too high cost of an artificial heart (Van 1995), and that smart card technology saves the health system money by preventing too common hospital admissions due to elderly patients receiving numerous prescriptions that can cause health problems when ingested together (Walker 1995). One article talks about a GPA tracking device for asylum seekers and immigrants, whereby they mention that the system they bought from the UK was originally used there to track elderly people with dementia (Chase 2013). Another article reports that changes in the Canadian patent law might endanger provincial drug plans for the elderly (Montgomery 1983). Various government officials are quoted with the prediction that healthcare will eat up 50% (Heyman 2002) to nearly 100% of provincial budgets (Ohler 2001) by 2012-2022 due to the aging population and cost of technology. Biotechnology and pharmaceuticals are seen as

---

5 | Before 1980 n= 0; 1980- 1989 n=1; 1990- 1999 n= 3; 2000- 2009 n=1; 2010-2014 n=0.

the top-performing industries because of the aging population (Yedlin 2001). Articles cover politicians' views that the cost of technology and the aging population demands a health care reform in Alberta (Heyman 2002) and reform of the Canada Health Act (May 2000), as well as the fact that a shift to private systems might take place (Cattaneo 1995). However, the coverage of the topic of cost related to technology dropped sharply after 2002, with the majority of articles covering cost deriving from the 1980's and 1990's.

To compare the newspaper coverage of cost with how cost is covered in the two Canadian reports, the 2009 Canadian report mentions cost n=94 throughout its report. However, it does only mention cost twice in relation to technology, namely that telehealth and electronic health records lower the cost of the health system. No negative cost impact of technologies is mentioned. The 2013 Alberta report mentions cost n=23 in general. However, in relation to technology, it only mentions cost twice, once talking about the decreasing cost effect of technologies: "The introduction of these technologies [medication management technology, mobility technology will be of tremendous benefit to the health-care system by reducing strains on the workforce; reducing serious health incidents through prevention and assistance; lowering costs by reducing visits to emergency departments; increasing health system capacity by safely supporting seniors at home; reducing demand on continuing care; and, reducing long-term stays in acute care."(:63) At the same time, the report features a quote by the Canadian Foundation for Healthcare Improvement that sees technological innovation as a main cost driver: "Some of the best research shows that, although health-care costs will begin to rise as baby-boomers age, the impact will be modest in comparison to that of other cost drivers, such as inflation and technological innovation." (:10) The 2002 Madrid action plan mentions cost n=14, but does not mention cost with regard to technological advancements. In other words both reports have cost as a main theme but not in relation to technology.

Specific Theme: Impact of Technology and Problem Narratives around Technology

As to impact of technology n=12 articles in the *Calgary Herald* and *The Globe and Mail* state negative impact of technology on the healthcare system and n=6 articles state positive aspects on the healthcare system.

As to the use of the term 'problem' most articles saw technology as tool to fix a problem but did not cover technology as the 'problem'. The findings in the newspapers is reflected in the two Canadian reports the 2009 Canadian report and the 2013 Alberta report and the 2002 Madrid action plan whereby technologies are mostly covered positively and not in a 'causing problem narrative'.

We posit that the reader does not receive a realistic picture of the impact of technologies as it is mostly covered in a technology to the rescue narrative.

As to the negative impact of technology (n=7) on the healthcare system the *Calgary Herald* had one article in 2013, one article in 2006, one in 2005 one in 2002 and the rest before then mentioning among others that "parents and children have to be more self-reliant if Albertans are going to have the money to provide the sort of high-tech medicine people" (Walker 1996: B3). As to *The Globe and Mail,* all n=5 articles highlighting the negative impact on the health care system were before 2001, with one article giving voice to the sentiment that the aging are not to be blamed, but increased service intensity and costlier technology instead (*The Globe and Mail* 1993). N=3 articles state that Canadians have poor access to medical/health technologies. One article wrote about the need for reasonable standard of service and maintaining a fiscal accountability over the system (Kennedy 2000).

As to the positive impact of technology on the healthcare system, the newest article in the *Calgary Herald* came from 2001, covering tele-health; one mentioned the cost savings through information technology and one focused on the saving of time through Internet and phone based technologies. Other themes present that are linked to the healthcare system but do not directly mention the healthcare system were that nurses need to upgrade their knowledge, that Canada is not great in providing access to healthcare technologies, that Canada lacks behind, that technology assessment is needed, and that wireless technology puts pressure on healthcare systems to organize patient data. As to *The Globe and Mail,* one article in 2012 and two in 2011 highlight the positive impact of technology (information technology) on the healthcare system (saving money and time).

Interestingly, within the three newspapers (*The Globe and Mail, Calgary Herald, National Post*) only six use the term "problem" in relation to technology and aging and the elderly. A 2002 *Calgary Herald* article described the design process of safer apartments for the frail elderly and which problems the elderly face could be solved through technology (Bertrand 2003). In *The Globe and Mail,* five articles were found: themes mentioned were that one can focus on smaller problems that can be dealt with through technology in a shorter framework; that Canadians do not plan how they want to die at a time technology can keep one alive; two articles focused on problems technologies could tackle; and one mentioned diffusion of the product from being used by people with a certain problem to the usage by the general public.

As to the Canadian and the Alberta report and the Madrid action plan all three mention technologies, mostly positively with little to no negative mentioning. The 2013 Alberta report for example, states that technologies "will enable many of us to stay in our homes longer and maintain greater control over our own health" (17), that "technology also holds the promise of supporting greater safety, autonomy and personal choice, especially for those of us who want to stay in our own homes for as long as possible" (18) and that "manage personal risks, reduce social isolation and carry out the tasks of everyday living. These technologies can also serve to bring peace of mind to families and caregivers, not to mention reduce demands on the health-care system (18).

Missing Theme: Socially Disadvantaged Groups

Another finding in our study is the lack of newspaper coverage of technologies in relation to aging of people from socially disadvantaged groups such as indigenous people (n=0), disabled people (n=1) or immigrants (n=0), as well as a lack of a gendered analysis (women only mentioned n=1). This seems to fit with the Canadian and Alberta report and the Madrid action plan. Although socially disadvantaged groups are covered in these three documents they are not mentioned in regards to technologies. The newspapers analysed, the Canadian and Alberta report and the Madrid action plan leave the impression that the technology and aging interaction is a homogeneous process; however, technology has different impacts on the aging experience of people from disadvantaged groups (for example for the case of people aging with a disability see (Wolbring/Lashewicz 2014). The lack of mention of socially disadvantaged groups around technologies and aging in the newspapers might be explainable as socially disadvantaged groups are not much mentioned in relation to many aspects of aging in the newspapers (Abdullah/Wolbring 2013), which in turn might be understandable as newspapers only tailor to their majority readership and what they want to hear (Arceneaux 2011; Gentzkow/Shapiro 2010; Stroud/Muddiman 2013). However, socially disadvantaged groups are covered in the Canadian and Alberta report and the Madrid action plan and it is less understandable why no linkage is made to technologies in these three documents. As such, the newspapers and for that matter the Canadian and Alberta report and the Madrid action plan set a discussion agenda around technology and aging that does not include socially disadvantaged groups and add to an uneven discussion around technologies and aging as it relates to socially disadvantaged groups.

Missing Theme: Ethics

Ethics hardly coincided with technologies related to the elderly/aging. Articles questioned the focus on the autonomy of robots and liability (n=2), what the public system should pay for, how to treat the elderly, whether to use IVF at a certain age, and issues related to life extension and high-technology medicine. Furthermore, although some articles mention the terms policy and governance – some even in the same sentence as technology – not one covers technology policy or technology governance in relation to aging or the elderly. Interestingly, both the Canadian and the Alberta report do not use the terms ethics or governance at all, and the 2002 Madrid action plan uses ethics only in relation to issues other than technologies.

This does not reflect the academic literature where ethics is an integral aspect of the discussion around many technologies that are applied to aging and the elderly. However, this finding might reflect that the public does not use the term ethics or thinks about 'ethical' constructs in abstract ways (see, for example, how parents of children with cognitive differences frame their concerns about cognitive enhancement (Ball/Wolbring 2014). Indeed Susan Sherwin, a highly influential bioethicist, believes that ethicists do not have the intellectual tools to really influence discourses (Sherwin 2011). Sherwin's identified problem might be due to the fact that ethical reasoning might not be the crucial factor, but the utility of a given ability expectation (Wolbring 2012). This, in other words, means that if we have certain ability expectations – such as independence or self-sufficiency – one will make an argument that justifies this ability expectation whereby the 'lay public' will use arguments understood by people they interact with. As stated elsewhere by the author, "this does not mean that ethical theories are not employed, but that if a certain ethical theory is employed and convinces people, it is because the ethical theory allows for living out certain ability expectations; for example, people who believe in the ability to maximize one's self interest might be open to someone promoting certain actions using the ethical theory of psychological/ethical egoism." (Wolbring 2012: 299)

Missing Theme: Ageism

Ageism is only once mentioned in the newspapers in relation to technology. As to the Canadian and the Alberta report and the Madrid action plan, the Alberta report mentioned ageism three times with regard to aging/elderly, but not once in relation to technologies. The Canadian report has ageism as the title of chapter 1: however, no linkage between technology and ageism is made whether in a positive or negative way. The 2002 Madrid action plan mentions ageism only once, asking the media to avoid ageism and present a more positive image of older persons.

The lack of coverage of ageism linked to technology in the newspapers and the Canadian and the Alberta report and the Madrid action plan again does not reflect data of many academic studies that show ageism exhibited in technology discourses (e.g. (Cutler 2005)).

In general, the results from our study suggest that newspapers are not a source to use if one wants to learn about the interaction between technology and the elderly/aging. However, the question is: how one does learn about it?

4. CONCLUSION: MISSING ITS MISSION

We conclude that the reader is not educated on key policy developments that cover aging and technology and does not obtain a useful picture of the interaction of technology and aging/elderly. The coverage does not fulfill the described role of newspapers, such as enabling social participation (Nord 1988). We posit that the information given is much too limited to allow the reader to be an informed discussant or even to know what to discuss. Newspapers do not increase literacy of the public on the topic, a problem that is not just linked to aging and technology, but that exists in other contexts as well (Wolbring et al. 2012), while newspapers are seen to set the discussion agenda (Wallack 1990). We posit that the coverage suggests an agenda that minimizes the angle of socially disadvantaged groups such as immigrants, disabled people or indigenous people. This makes the coverage problematic for disadvantaged social groups and ignores the attention that the three policy documents covered in our study state disadvantaged groups should receive. The reader also does not learn about the gendered aspect of technology, which is also a problematic issue.

Furthermore, the newspapers do not cover policy documents such as the 2009 Canadian report, the 2013 Alberta report or the Madrid action plan. This finding is similar to another study that found that key documents about active aging, such as the 2002 *World Health Organization report Active Ageing: a policy framework* and the 2010 *Toronto Charter for Physical Activity: A Global Call for Action,* are not covered by the newspapers represented in the *Canadian newsstand complete* database either (Abdullah/Wolbring 2013). This, we posit, is a problem if the public is to be informed in a useful way on a topic that demands that the public should be alerted on policy documents.

The 2009 Canadian report sees a need for policy makers to be better informed on aging:

"Policy-makers need to base their decisions on sound evidence and a grounded understanding of the many ways people age. This will require ongoing, longitudinal research to understand the process of aging and the complex ways in which economic, social and health factors affect in which people age well or not. Seizing the opportunity of an aging population will also require a better understanding of how technological advances can be used to improve the quality of life of Canadians and to make the most efficient use of limited human resources." (155)

We propose that policy makers do not gain this knowledge through newspapers but have to access this knowledge through other sources. Furthermore, the same knowledge is needed by the public at large, but the question is where they would obtain such knowledge if newspapers are not the source of it. Policy makers have access to sources – such as academic databases – that the public does not: this suggest that there is a knowledge access gap between policy makers and the public, which we contend is problematic given the narrative that the public should be part of technology governance processes.

Although there are many problems with the coverage, the question is what to do? Are newspapers really able to regain the role they were supposed to play? Can other media such as social media fulfil this role? The answers to these questions are less clear. Social Media might allow people who are already interested to be part of, for example, certain facebook pages. However, to obtain the knowledge which might lead one to become active, social media might be too overwhelming to figure out what the issue is. Another possibility is to get involved locally on the topic face to face, although one must first be aware that there is an issue, an awareness that one might only obtain through direct experience of a problem within one's social circle. This approach also assumes one can access the local networks: this is more than problematic for many people with disabilities, including elderly with disabilities, whether they live at home or in other places such as senior homes.

REFERENCES

Abdullah, B./Wolbring, G. (2013): "Analysis of Newspaper Coverage of Active Aging through the Lens of the 2002 World Health Organization Active Ageing Report: A Policy Framework and the 2010 Toronto Charter for Physical Activity: A Global Call for Action." In: International Journal of Environmental Research and Public Health 10/12, pp. 6799-6819.

Angus, J./Reeve, P. (2006): "Ageism: A threat to "aging well" in the 21st century." In: Journal of Applied Gerontology 25/2, pp. 137-152.

Arceneaux, K., (2011): Niche News: The Politics of News Choice, by Natalie Jomini Stroud, New York: Oxford University Press.

Armstrong, J. (2002): "Canada is 30 million, but will that last?" In: The Globe and Mail. Montreal, March 13, pp. A1.

Ball, N./Wolbring, G. (2014): "Cognitive Enhancement: Perceptions Among Parents of Children with Disabilities." In: Neuroethics 7/3, pp. 345-364.

Berridge, C./Furseth, P.I./Cuthbertson, R./Demello, S. (2014): "Technology-Based Innovation for Independent Living: Policy and Innovation in the United Kingdom, Scandinavia, and the United States." In: Journal of Aging & Social Policy 26/3, pp. 213-28.

Bertrand, J.-F. (2003): "Creative solutions ease tasks for elderly." In: Calgary Herald. Calgary, June 28, pp. E2.

Binstock, R.H. (1992): "The oldest old and 'intergenerational equity'." In: The oldest old, Suzman, M./ Willis, D. (eds.), Oxford: Oxford University Press, pp. 394-417.

Buchholz, M./Bynum, J.E. (1982): "Newspaper presentation of America's aged: A content analysis of image and role." In: The Gerontologist 22/1, pp. 83-88.

Special Committee on Aging, Parliament of Canada (2009): Canada's aging population: Seizing the opportunity, March 6, 2015 (http://www.parl.gc.ca/Content/SEN/Committee/402/agei/rep/AgingFinalReport-e.pdf)

Cattaneo, C. (1995): City leads private health trend. In: Calgary Herald. Calgary, November 3, pp. A1.

Chan, J./Chuang, L./Yeh, H. (2014): "Key factors in the design of next generation portable supporting devices for elderly adults." In: Gerontechnology 13/2, pp. 137-138.

Chase, S. (2013): "Border agents plan to buy GPS bracelets." In: The Globe and Mail. Montreal, June 27, pp. A4.

Cheung, J./Wolbring, G. (2015): "Analysis of the Science and Technology Narrative within Organ Donation and Transplantation Coverage in Canadian Newspapers." In: Technologies 3/2, pp. 74-93.

Chief Medical Officer of Health Alberta (2013): "Let's talk about aging: aging well in Alberta" March 6, 2015 (http://www.health.alberta.ca/documents/CMOH-Aging-In-Alberta-Report-2013.pdf).

Cutler, S.J. (2005): "Ageism and technology." In: Generations 29/3, pp. 67-72.

Fealy, G./McNamara, M./Treacy, M.P./Lyons, I. (2012): "Constructing ageing and age identities: a case study of newspaper discourses." In: Ageing and Society 32/1, pp. 85.

Flandorfer, P. (2012): "Population Ageing and Socially Assistive Robots for Elderly Persons: The Importance of Sociodemographic Factors for User Acceptance." In: International Journal of Population Research 2012 (2012) Article ID 829835, 13 pages http://dx.doi.org/10.1155/2012/829835.

Gentzkow, M./Shapiro, J.M. (2010): "What drives media slant? Evidence from US daily newspapers." In: Econometrica 78/1, pp. 35-71.

Gusmano, M.K./Allin, S. (2014): "Framing the issue of ageing and health care spending in Canada, the United Kingdom and the United States." In: Health economics, policy and law 9/3, pp. 313-328.

Heath, J. (2013): "The Structure of Intergenerational Cooperation." In: Philosophy & Public Affairs 41/1, pp. 31-66.

Heerink, M./Krose, B./Evers, V./Wielinga, B. (2010): "Assessing acceptance of assistive social agent technology by older adults: The almere model." In: International journal of social robotics 2/4, pp. 361-375.

Heyman, D. (2002): "Mazankowski fends off critics: Lobby group says data 'misconstrued'." In: Calgary Herald. Calgary, May 14, pp. A6.

Hubbard, R.E./Eeles, E.M./Fay, S./Rockwood, K. (2009): "Attitudes to aging: a comparison of obituaries in Canada and the UK." In: International Psychogeriatrics 21/4, pp. 787.

Kang, S./Gearhart, S./Bae, H.-S. (2010): "Coverage of Alzheimer's disease from 1984 to 2008 in television news and information talk shows in the United States: an analysis of news framing." In: American Journal of Alzheimer's Disease and Other Dementias 25/8, pp. 687-697.

Kennedy, M. (2000): "Nfld. premier questions medicare's vitals." In: Calgary Herald. Calgary, May 2, pp. A5.

Mastin, T./Choi, J./Barboza, G.E./Post, L. (2007): "Newspapers' framing of elder abuse: It's not a family affair." In: Journalism & Mass Communication Quarterly 84/4, pp. 777-79.4

Matcha, D.A./Sessing-Matcha, B.A. (2007): "A Comparison of American and European Newspaper Coverage of the Eldery." In: Hallym International Journal of Aging 9/2, pp. 77-88.

May, H. (2000): "Private health care called the answer." In: Calgary Herald. Calgary, June 24,pp. A19.

Miller, E.A./Tyler, D.A./Rozanova, J./Mor, V. (2012): "National Newspaper Portrayal of US Nursing Homes: Periodic Treatment of Topic and Tone." In: Milbank Quarterly 90/4, pp. 725-761.

Molnar, F.J./Man-Son-Hing, M./Dalziel, W.B./Mitchell, S.L./Power, B.E./Byszewski, A.M./John, P.S. (1999): "Assessing the quality of newspaper medical advice columns for elderly readers." In: Canadian Medical Association Journal 161/4, pp. 393-395.

Montgomery, C. (1983): "Drug costs at stake in patent law battle." In: The Globe and Mail, Montreal, November 12, pp. P16.

Nord, D.P. (1988): "A Republican Literature: A Study of Magazine Reading and Readers in Late Eighteenth-Century New York." In: American Quarterly 40/1, pp. 42-64.

Ohler, S. (2001): "Pannu writes prescription funding plan: Full coverage for seniors." In: Calgary Herald. Calgary, March 2, pp. A18.

Peek, S./Wouters, E.J./van Hoof, J./Luijkx, K.G./Boeije, H.R./Vrijhoef, H.J. (2014): "Factors influencing acceptance of technology for aging in place: A systematic review." In: International Journal of Medical Informatics 83/4, pp. 235-248.

Peine, A./Rollwagen, I./Neven, L. (2014): "The rise of the 'innosumer'— Rethinking older technology users." In: Technological Forecasting and Social Change 82, pp. 199-214.

Powell, M. (2013): What Do Newspapers Say About Older Adults in the Workforce?: Ageism and Mistreatment of Older Workers, Brownell, P./Kelly, J. (eds.), New York:Springer, pp. 49-67.

Rozanova, J. (2010): "Discourse of successful aging in The Globe & Mail: Insights from critical gerontology." In: Journal of Aging Studies 24/4, pp. 213-222.

Second World Assembly on Ageing (2002): "Madrid International Plan of Action on Ageing", March 6, 2015 (http://undesadspd.org/Portals/0/ageing/documents/Fulltext-E.pdf).

Sherwin, S. (2011): "Looking backwards, looking forward: Hopesfor bioethic's next twenty five years." In: Bioethics 25/2, pp. 75-82.

Speirs, R. (1983): "Ontario Tory policy makers favor leaner social services." In: The Globe and Mail. Montreal, September 26, pp. P1.

Stroud, N.J./Muddiman, A. (2013): "Selective Exposure, Tolerance, and Satirical News." In: International Journal of Public Opinion Research Online 25/3, pp. 271-290

The Globe and Mail (1993): "AGING DEMOGRAPHICS Social spending destined to rise." In: The Globe and Mail. Montreal, September 17, pp. C1.

The Globe and Mail (1989): "Vencap Medical Ventures." In: The Globe and Mail. Montreal, March 21, pp. B10.

Tynedal, J./Wolbring, G. (2013): "Paralympics and Its Athletes Through the Lens of the New York Times." In: Sports 1/1, pp. 13-36.

United Nations Department of Economic and Social Affairs Population Division (2012): "Population ageing and development: Ten years after Madrid." March 6, 2015 (http://www.un.org/en/development/desa/population/publications/pdf/popfacts/popfacts_2012-4.pdf).

United Nations Department of Economic and Social Affairs Population Division (2001): "World Population Ageing: 1950-2050." March 6, 2015 (http://www.un.org/esa/population/publications/worldageing19502050/).

Van, J. (1995): "Heart pumps make comeback." In: Calgary Herald. Calgary, November 23, pp. B9.

Walker, R. (1996): "Patients need to toughen up." In: Calgary Herald. Calgary, March 4, pp. B3.

Walker, R. (1995): "'Smart cards' to cut health costs Series: Taking the pulse: Health-care revolution." In: Calgary Herald. Calgary, Jue 15, pp. A2.

Wallack, L. (1990): "Mass media and health promotion: Promise, problem, and challenge." In: Atkins, C./Wallack, L. (eds.), Mass communication and public health: Complexities and conflicts, Newbury Park: Sage, pp. 41-53.

Weicht, B. (2013): "The making of 'the elderly': Constructing the subject of care." In: Journal of Aging Studies 27/2, pp. 188-197.

Weinstock Netanel, N. (1996): "Copyright and a Democratic Civil Society." In: Yale Law Journal 106/2, pp. 292-392.

Wolbring, G. (2012): "Ethical Theories and Discourses through an Ability Expectations and Ableism Lens: The Case of Enhancement and Global Regulation." In: Asian Bioethics Review 4/4, pp. 293-309.

Wolbring, G./Lashewicz, B. (2014): "Home care technology through an ability expectation lens." In: Journal of medical Internet research 16/6, pp. e155.

Wolbring, G./Leopatra, V./Yumakulov, S. (2012): "Information Flow and Health Policy Literacy: The Role of the Media." In: Information 3/3, pp. 391-402.

Wolbring, G./Yumakulov, S. (2014): "Social Robots: Views of Staff of a Disability Service Organization." In: International journal of social robotics 6/3, pp. 457-468.

World Health Organization (2002): Active ageing: A policy framework, March 6, 2015 (http://whqlibdoc.who.int/hq/2002/who_nmh_nph_02.8.pdf).

Yedlin, D. (2001): "Investors urged to show courage." In: Calgary Herald. Calgary, February 23, pp. A1.

Yumakulov, S./Yergens, D./Wolbring, G. (2012): "Imagery of people with disabilities within social robotics research." In: Social Robotics: 4th International Conference, ICSR 2012, Chengdu, China, October 29-31, 2012. Proceedings , Lecture Notes in Computer Science, New York: Springer, pp. 168-177.

Towards an Ageless Society: Assessing a Transhumanist Programme

MARTIN SAND AND KARIN JONGSMA

1. TRANSHUMANISM: AN EXTREME TECHNOLOGICAL VISION

In recent days, a neglected factor in technological development came to the fore in the fields of Technology Assessment (TA) and Science and Technologies Studies (STS): visions as a medium to communicate uncertain technological futures (Grunwald 2009, 2012)[1]. Outside the scientific community there is also an increasing demand for orientation for technological futures. Many societal agents such as citizens, politicians and entrepreneurs are interested in foreseeing consequences of emerging technologies (Grunwald 2013). The high promises as much as the apocalyptic scenarios of emerging

---

[1] | Though there have been a few studies on the functioning of "Leitbilder" and visions in shaping technologies in the early and middle nineties (Dierkes/Hoffmann/Marz 1992; van Lente 1993), TA became increasingly aware of those visions just since the beginning of the nineties. The cause for this awareness lies in the development of the societal discourse on the Convergence of Nano-, Bio-, Cogno- and Information Sciences (NBIC) to improve human capacities (Roco/Bainbridge 2003; Coenen 2009). Even though the Transhumanistic vision is much older than the NBIC debate (Heil 2010), it has its central reference to the potential of Nanotechnology for the improvement of the human condition as well as many similarities with the vision of NBIC. Besides this, personal overlapping between the two discourses binds these historical strings together (Coenen 2006). The above quoted William S. Bainbridge for instance, famous for his contribution in the NBIC debate, has recently published an article on the idea of „Transvatars" in the „Transhumanist Reader" (Bainbridge 2013).

technologies raise public attention, give reasons for hope, but also create anxiety and fear (Simakova/Coenen 2013). Uncertainties at the early stage of technological development create a demand for orientation and assessment of technological visions. Hence, stakeholders and politicians want to identify possible requirements for regulations (e.g. bio-safety), for chances of action (e.g. investing) or they just want to be informed about the likelihood of certain promises made in the field of emerging technologies (Grunwald 2013: 27). Technological visions do not only influence the general public as a passive recipient; they can be considered as narratives that also guide scientists and engineers in a certain direction of technological development. This guidance subsequently shapes the design of technologies: hence, visions are considered as an important parameter in technological development.

Transhumanism is an extreme technological vision, which clearly distinguishes itself from utopian narratives such as Thomas More's *Utopia*[2] in a few respects. In contrast to such classic utopian narratives, Transhumanism does not provide a general critique of the contemporary social and institutional order. Instead of criticizing the liberal market or the post-democratic political system for instance, Transhumanism accepts and seeks to sustain this system with its fostered values of efficiency and accomplishment (Saage 2006, 2007). The object of improvement is the individual and not the social system that may have developed the pressure on individuals that Transhumanists want to master. A further major difference between utopias and the vision "Transhumanism" is the fact that the envisioned change should be accomplished by technological means. Transhumanists consider technology as the appropriate means to deal with fundamental societal issues and to fulfil deep human desires. While former utopias considered a change of the reigning political and economic orders or education as promising means, Transhumanists prioritize technology as a fundament for improvement. The vision of Transhumanism appears in various forms, all of which have in common that they envision a future that brings about a decline of our physical and mental

---

2 | Thomas More's *Utopia* is seen as the prototype of a utopia and one of the constitutive literary documents of the modern age (Nipperdey 1975). The book contains a fictive dialogue in which criminal behaviour is discussed. One of the dialogue participants concludes that a fundamental change of the social order is necessary to overcome such behaviour. It is suggested that such a utopia is a possible and ideal world. Utopia is pictured as a universal rearrangement of the political, social and institutional order (Nipperdey 1975: 118).

boundaries as compared to today. The unpleasantness of all physical constraints – including diseases, senescence and death – should be overcome: Transhumanists seek to expand humanity into space. Transhumanist Max More summarizes these goals in his overview on the "Philosophy of Transhumanism":

Becoming posthuman means exceeding the limitations that define the less desirable aspects of the 'human condition'. Posthuman beings would no longer suffer from disease, aging, and inevitable death (but they are likely to face other challenges). They would have vastly greater physical capability and freedom of form – often referred to as "morphological freedom" [...]. Posthumans would also have much greater cognitive capabilities, and more refined emotions (more joy, less anger, or whatever changes each individual prefers). Transhumanists typically look to expand the range of possible future environments for posthuman life, including space colonization and the creation of rich virtual worlds. (More 2013a: 4)

The *Transhumanist Declaration* also expresses these prophetic claims and the trust in the techno-scientific progress in a typically positive tone:

Humanity stands to be profoundly affected by science and technology in the future. We envision the possibility of broadening human potential by overcoming aging, cognitive shortcomings, involuntary suffering, and our confinement to planet Earth. (Various 2013: 54)

While one could be inclined to regard Transhumanist visions as far-fetched speculations about uncertain futures and therefore neglect their significance and meaningfulness, it must be emphasised that visionary practices have a specific impact on the development of technologies in the present. Visionary narratives narrow the scope of possible paths in which we could develop new technologies. The Transhumanist vision regards technology as the fundament for future welfare: this leads to a distraction of the public discourse and neglects the opportunities of non-technological improvement of our society. Furthermore, the involvement of Transhumanists in concrete scientific and engineering enterprises shows that technological visions do not float in an empty space, but rather guide concrete contemporary business and engineering activities[3]. Recently, Patrick McCray described the set of activities that many promoters of extreme technological visions are engaged in as "visioneering" practises (McCray 2013). Visioneering means the fostering of the visionary community, advocating for technological enterprises and the visionary agenda publicly, doing research to realise the vision, and allocating funding for these projects. Visioneers, such as Transhumanists, function as important actors in the social construction of technologies. The impact of their activities and the dangers emerging from it become obvious in the personal and conceptual connection between Transhumanism, biogerontology and anti-ageing medicine, which is manifested, for instance, in the person of Aubrey de Grey[4]. In the following section, we will discuss the advocates of the Transhumanist movement and deal with the Transhumanist vision of ageing and its technological claims.

2. Transhumanist Views on Ageing

Transhumanists claim that ageing leads to unnecessary suffering and death. Not without a reason, the Transhumanist Julian Huxley begun a list of desirable goals that could soon be achieved with a long and healthy life through scientific progress (Huxley 1931: 5–6). Just like Julian Huxley, Nick Bostrom assumes that almost everyone would choose a radically prolonged life as long as one could undergo this extended lifespan in an acceptable healthy condition (Bostrom 2013: 33). The claim of the Transhumanists is, therefore, not only to extend the lifespan but also to extend the productive years of life – the "healthspan", as Bostrom names it. He argues that this social desire is visible in many contemporary technologies such as the airbag. The development of these technologies is driven, as Bostrom says, by the wish to avoid unnecessary and premature deaths and injuries that subsequently lead to dying. The increase in the lifespan of the western civilization in the past century

---

3 | Since 2012 the leading Transhumanist Ray Kurzweil is director of engineering at *Google*. He is a good example of an active visioneer.

4 | Aubrey de Grey is a famous figure of the Transhumanist movement. He is a fellow of the Transhumanist Institute for Ethics and Emerging Technologies and Chief Science Officer of the SENS Research Foundation. The foundation is dedicated to the research on ageing and had a total income of 15 Million $ through personal donations of Aubrey de Grey in 2012 (SENS Foundation 2013).

gives us evidence for the efficiency of the technological mastering of ageing and dying. Bostrom has no doubt that further progresses in the intervention to the ageing process are likely and soon to appear if the right decisions are made nowadays. In his famous, original and lucid article, "The fable of the dragon", he attacks conservative bioethicists who deny the desirability of an eternal life and promote cautiousness about the side-effects of technologies that intervene into the process of ageing and dying (Bostrom 2005). The moral of the fable is that such a conservative attitude delays the mastering of the ageing process and subsequently costs unnecessary lives. Bostrom suggests that a more liberal research and technology agenda could save those lives. He pleas for a shift in the paradigm of technological development from the "precautionary" to the "proactionary principle" (More 2013b). The fable shows that the reign of the dragon, which stands for human mortality, is for Bostrom not a law of nature that cannot be influenced. The accusations that conservative bioethicists and other parties oppose the liberal development of technology, delay the progresses in life-prolongation, and are therefore responsible for unnecessary and premature death, are common in Transhumanism. In this debate, ageing is considered as the biological cause of diseases and premature death. The fight against ageing is therefore a fight against the cause for human mortality. The quest is to intervene in the ageing process in order to overcome it. Eric Drexler, the author of the "Engines of Creation", published in 1986, considered cell-repair machines based on nanotechnology as the appropriate tool to intervene in the ageing process. Drexler's promises with regard to Nanotechnology supported the strong expectations many Transhumanists associate with Nanotechnology. Drexler assumes that ageing is like other biological processes – natural, though not less undesirable:

> Aging is natural, but so were the smallpox and our efforts to prevent it. We have conquered smallpox, and it seems that we will conquer aging. [...] Still, researchers have made progress toward understanding and slowing the aging process. They have identified some of its causes, such as uncontrolled cross-linking. They have devised partial treatments, such as antioxidants and free-radical inhibitors. [...] With cell repair machines, however, the potential for life extension becomes clear. They will be able to repair cells so long as their distinctive structures remain intact, and will be able to replace cells that have been destroyed. Either way, they will restore health. Aging is fundamentally no different from any other physical disorder; [...]. (Drexler 1986: 114–115)

This quote contains many aspects that are characteristic of the Transhumanist understanding of ageing, which reappears in many of their writings. A very typical characteristic of Drexler's way of reasoning in this quote, is that he considers ageing as a biological process; also typical in this quote is the claim that science has already made a huge progress in explaining these processes. Furthermore, his phrasing of ageing by using the military metaphor of "conquering ageing" suggests that society faces an opponent worth fighting, a truly evil enemy. Implied in this use of language is a personalisation of ageing as an entity that somehow has person-like attributes such as intentions or a will (to destroy humans): these metaphors reappear in many writings of Transhumanists and certainly develop an immense rhetoric power. He continues with another negatively phrased comparison: ageing as being similar to the unpleasantness of smallpox. Moreover, Drexler thinks that there are, or soon will be, technological ways to intervene in the ageing process and that we will and should apply these technologies: he envisioned the so called "Nanobots", based on Nanotechnology, to cure the cellular damage at its roots (Nerlich 2005). Finally, he considers ageing as a physical disorder, a phenomenon that does not only lead to diseases, but is a disease in itself.

Although Drexler's paradigmatic viewpoint on aging has many similarities with Transhumanism, it shows a difference to technological interventions that are currently summarized as anti-ageing technologies (Ehni 2014: 49). Transhumanists seek to completely reverse the ageing process: while anti-ageing medicine promises to undo the externally visible effects of ageing such as wrinkles, Transhumanists want to treat the roots of ageing, the biological process itself and not only its symptoms. However, the practises that are subsumed as anti-ageing medicine cover a wide range of scientific focuses and approaches (Everts Mykytyn 2010: 183). At first sight, it might seem that restoring health and fighting diseases is not a completely new claim: the example of Julian Huxley and the idea of the "Fountain of Youth" give evidence for a rather long tradition for this desire (Schade-Tholen/Franke 1998; Gruman 1966). The density and extent in which biomedical means are increasingly promoted as promising tools to intervene in the ageing process is, however, a historical novelty (Schermer/Pinxten 2013; Everts Mykytyn 2010). New is the way in which scientific progresses in the field are presented. Many proponents of ageing intervention technologies for instance prompt that the question is not "if", but rather "when" and "how" we will intervene. This suggests that we are dealing with a new quality of expectations (Achenbaum 2005; Everts Mykytyn 2010).

Besides Nick Bostrom and Eric Drexler, Aubrey de Grey is one of the most famous and criticized advocates of reversing the ageing process. In many publications he promotes society's war against ageing (Grey 2004): he co-founded the "Strategies for Engineered Negligible Senescence" (SENS) research foundation in 2009 and advertises it in many publications. By framing the SENS enterprise as an "engineered negligible senescence", this foundation seeks to distance itself from competing anti-ageing approaches such as face-lifting and other "surface interventions" in the ageing process. The metaphor of the bodily machine which is common in Transhumanism reappears in this framing: restoring the "normal" condition means repairing and this is usually done by engineers. De Grey's perspective on ageing has apparent similarities with Drexler's:

'In the context of discussing interventions, ageing can be defined as the lifelong accumulation of various intrinsic side effects of normal metabolic processes, which ultimately reach an abundance that disrupts metabolism and causes severe dysfunction of tissues and the whole organism. Some aspects of this dysfunction are classified as age-related diseases, and some less specifically as 'frailty', but their common cause is the accumulation of damaging metabolic side effects.' (Grey 2005: 49)

This quote shows that he regards ageing as the process of ongoing damage to the body. A strong normative wording describes the changes in the metabolic system and bodily functions of ageing people: the concepts 'dysfunction' and 'damaging' frame ageing negatively, as if the whole ageing body is a broken or dysfunctional machine. In his book *"Ending Aging: The Rejuvenation Breakthroughs That Could Reverse Human Aging in Our Lifetime"*, he writes that his research, and actually the whole enterprise of anti-ageing research, is about the abolishment of the unbearable suffering and pain that is caused by ageing (Grey/Rae/Burgermeister 2010: 16). In the book he argues that the development of age-reversing technologies is equally important to prior developments in medicine, like the development of penicillin. The fact that we regulate nicotine abuse by banning smokers from buildings and campaigning against smoking shows that society is interested in prolonging the lifespan in a healthy condition. Furthermore, the claims he makes rest mainly on findings in research on fruit flies and rodents (Rose 2004, 2013); in some laboratory experiments, caloric restriction has caused to slow down the ageing process in these animals. De Grey presents the assumption that responsible genetic components for the ageing process have already successfully been identified and manipulated (Grey 2005: 49). Aside philosophical doubts about the individual desirability of a radically prolonged live (Williams 1993; Ehni 2009) and concerns about the impact of a ubiquitous intervention into the ageing process with regard to global equality and possible generation conflicts (Fukuyama 2003: 57–71), many biogerontologists have also criticized de Grey's approach. The following chapter recaps the critique of the Transhumanist approach to ageing.

3. Some Remarks on the Reduction of Ageing and further Methodological Shortcomings

After having discussed how Transhumanists understand ageing, we will make four claims in this chapter: Firstly, in order to clarify the shortcomings of Transhumanist use of concepts related to ageing and health, some terminological remarks on the use of the concept "natural" and "disease" will be provided. Then we will come to the lack of scientific evidence for the claims of the Transhumanists: there are methodological doubts about the transferability from laboratory models to the human organism and a lack of verification of certain promoted interventions. Thirdly, we will explain why the concept of ageing cannot be understood solely as a biological process; it is necessary to describe ageing as a complex phenomenon that requires an interdisciplinary research approach to respond adequately to the challenges of an ageing population. The normative conclusions drawn on a one-dimensional approach to ageing are, therefore, inappropriate. We will end with arguing that the overwhelmingly positive picture of the development of future technologies is unjustified. Transhumanists neglect the fact that the technological progress is ambivalent, that is, that it contains positive and negative effects.

Conceptual Shortcomings

In the above quoted passages of Drexler and De Grey and in many other Transhumanist works, the concept of a "disease" describes an abnormal functioning of the body that is accompanied by shortcomings in well-being, ability, and productivity. As Anders Sandberg puts it with reference to Nanovisioneer Robert Freitas: "Disease is a failure of optimal functioning or desired functionality [...]" (Sandberg 2013: 63). The question of what a disease is and whether diseases are constructed or natural is too complex to be discussed here[5], but even if one would agree with the understanding of a disease as a malfunction, it seems that not all diseases lead to the same state of disability and lack of well-being. Ageing seems to lead to certain diseases; however, the identification of ageing with diseases is enriched with misunderstandings. If everything that leads to a certain state justifies the qualification of the cause as this state, then ageing could also be identified with dying. While one could skip the question of why one should draw such a conclusion[6], one may ask if this appropriately reflects the way we usually treat each other. Our conceptual pre-understanding of dying and ill persons is that such persons deserve special treatment and attention. An ill person has special needs and interests that we are obliged to take into account in our actions. A dying person appertains respect and consolation. If we age from the beginning of our lives, this ubiquitous attribution of disease does not represent the way we usually treat other people in our everyday lives, otherwise it would lead to the absurd accusation that all of us disobey the duties we actually have towards each other as ill people. Furthermore, from a pragmatic point of view we have to consider that this attribution makes dealing with diseases in regard to the law and, for instance, insurance practices unfeasible. A more or less clear demarcation between healthy and ill people is pragmatically necessary in various contexts. Finally, the distinction of age-caused and age-related diseases is substantive. The causes for many diseases that appear more frequently in old age are still unclear. It is reasonable to prioritize the term "age-related" or "age-associated disease" to the term "age-caused diseases" until there is more evidence for their source (Blumenthal 2003; Hayflick 2000).

Lack of Scientific Evidence

The presumed findings of Aubrey de Grey and his colleague Michael Rose are often criticized, also in the biogerontological community, as the quality of SENS research does apparently not fulfill scientific standards (Olshansky/Hayflick/Carnes 2002b). In their critique of the SENS approach, Hubert Warner and other gerontologists finally concluded their evaluation with a crushing statement:

---

5 | For an instructive overview see Murphy (2009).
6 | To illustrate this conflation of concept one may think about this analogous case: Is a storm the same as lightning just because it is almost always accompanied by lightning?

In our opinion, however, the items of the SENS programme in which de Grey expresses such blithe confidence are not yet sufficiently well formulated or justified to serve as a useful framework for scientific debate, let alone research. (Warner et al. 2005: 1008)

Rose and de Grey present selective results from evolutionary biology, cell biology and molecular biology, which they crudely merge to a comprehensive theory of ageing (Rose 2013: 200). The greatest success for an intervention into the ageing process to which they refer over and over again is caloric restriction. Already in 1934, C.M. McCay and Mary Crowell published a paper in which they presented success in prolonging the lifespan of mice through caloric restriction (McCay/Crowell 1934): Michael Rose has also managed to double the lifespan of fruit flies in his laboratories through caloric restriction (Rose 2004: 25). These results may impress at first sight, although evidence for an extrapolation from lower mammals and insects to human organisms has never been provided. It is doubtful whether such an extrapolation is defendable without any concrete investigation into the relation the relation between caloric restriction and the lifespan of human organisms (Sierra/Hadley/Suzman/Hodes 2009; Olshansky/Carnes 2013). De Grey mentions various further methods of intervention: he proposes, for example, that we need to rely on stem cell therapy and genetic engineering to stop the ageing process. He further suggests that small molecule drugs and somatic gene therapy could support this aim and subsequently even reverse the ageing process (Grey 2003: 930). Unfortunately, he does not elaborate on these technologies, how to create and use them. Damien Broderick, also a promoter of Transhumanism, points out that in 1999 Cynthia Kenyon and Herbert Boyer from the University of California successfully identified a gene that is responsible for the cell metabolism in *Caenorhabditis elegans*, a nematode (Broderick 2001: 39–41). By just depriving the worm from the gene, the lifespan could be doubled. Broderick mentions that researchers at Emory University have been likewise successful by adding two synthetic enzymes, superoxide dismutase and catalase, which combat oxidative stress. This research has also been carried out with *C. elegans* worms. While this sort of laboratory experiment has particularly been successful in identifying some factors of ageing in these lower model organisms, the transferability of these results is highly questionable. The *C. elegans* worm consists of about 960 body cells; the human body, in contrast, contains about hundred trillion (10^{14}) cells. None of the presented approaches has ever been proved to be primarily responsible for the human ageing process and neither has one of the technological interventions ever been tested in humans (Warner et al. 2005). Besides, the side-effects of these therapies for human organisms are unknown.

The Reductionist View on Ageing

Eric Drexler and de Grey assume that ageing is no different from any other physical disorder. One can wonder what this biological approach to ageing says about ageing in general. In the article *"The Problem of Theory in Gerontology Today"* from 2005, the authors argue that three main issues in contemporary gerontological research have been identified: biological and social processes of ageing, the aged themselves, and age as a dimension of structural and social organization (Bengtson/Putney/Johnson 2005). This list makes some relevant distinctions. It says that the process of ageing is biological *and* social, which is a crucial observation; furthermore, this approach points out that a precondition to investigate ageing is the notion of an aged person. Research in the psychology of ageing has shown that there is no measurable decline in the intelligence of elderly people, but rather an improvement of practical skills (Sternberg/Grigorenko 2005). Besides, there are several illustrative findings in the role of elderly persons of authority in the family bond, the age-related setting of new goals, and the effects of late career climaxes and retirement: these studies observe and investigate age-related social and psychological developments. Mental and social changes in late life are two of the numerous features that are part of the ageing process; if these sociological and psychological observations are to be taken into account to express a comprehensive theory of ageing, this requires an interdisciplinary approach from current gerontological research. The above mentioned gerontologists Bengtson, Putney and Johnson underline this requirement when they write:

The field of gerontology itself is in need of integration, because so many more factors are now recognized to be involved in human ageing. For the mountains of data to yield significant new insights, an integrating framework is essential. But this cannot be done without theories and concepts that are broader and more general in scope. This lack of integration in theories of ageing is also an artifact of disciplinary specialization. (Bengtson/Putney/Johnson 2005: 6)

By referring to this quote, one does not need to assume that it will remain impossible to explain social and cultural phenomena through biological theories, but this field of research is currently underdeveloped and does not do justice to the complex reality of human ageing. The Transhumanist 'biologisation of ageing' (Vincent 2013) is to be understood as a reduction of ageing. What does that mean? The conceptual question of what ageing is, is a metatheoretical one (Janich 2009). An answer to that question is, as mentioned above, a prerequisite in certain gerontological approaches. De Greys' analysis, for instance, starts with a preconception of ageing that is solely biological, and the research on fruit flies will not disclose any other social or cultural features of ageing. If the results of this research would suggest an answer to the question of what ageing is, the argumentation would be perfectly circular. So far, the biological concept of ageing thus fades out many relevant aspects of ageing that are observed by the humanities and the social sciences: it is therefore no wonder that Transhumanists conclusively value ageing negatively. An increase of well-being through a fulfillment of personal wishes, interests or an individual plan is also part of the ageing process, and the experiences we gather throughout our lives and the social contacts we foster partly constitute the value of our lives. This is the reason why we actually identify certain ages with specific features such as youthful carelessness[7].

Positive Picture of Technological Progress is Unjustified

Technological progress is ambivalent (Ropohl 1991; Lenk 1994): it is accompanied by many positive aspects such as an increasing mobility, for instance, but also with burdens for the environment and the climate. This seems to be a trivial observation yet it contrasts with the way Transhumanists present and frame our technological future. In Nick Bostrom's "Fable of the Dragon", for instance, one cannot find any remark that the accelerating 'wheel of invention' causes any negative effects, except that it costs money (Bostrom 2005: 276–277). The possibility that such progress harms the environment,

---

[7] | In Lewis Carroll's "Alice in Wonderland" a poem illustrates the attribution of age-related features humorously. An astonished teenager asks his father how he gained all his impressive skills: "In my youth," said his father, "*I took to the law, and argued each case with my wife; And the muscular strength which it gave to my jaw, has lasted the rest of my life.*" "You are old," said the youth, "*one would hardly suppose, that your eye was as steady as ever; Yet you balanced an eel on the end of your nose – What made you so awfully clever?*" (Carroll/Gardner/Tenniel 2000: 51)

persons or other beings that are involved in research trials, which may lead to even greater problems than in our current ageing society, is neglected. This shortcoming is part of a vision that is based on a naïve trust in the benefits of technological progress: therefore, we argue that the vision of Transhumanism is clearly only a one-sided view on technological interventions, lacking reasonable skepticism.

In the following concluding remarks we will come back to our point of departure: the Transhumanist vision and we will propose an alternative view concerning ageing and technological innovation.

4. CONCLUSIONS: RENOVATING THE VISION OF ANTI-AGEING

Transhumanists promote an extreme technological vision that receives increasing public attention. Charismatic characters such as Ray Kurzweil, Nick Bostrom and Aubrey de Grey – who have a scientific background and engage in the free economy, but also act with a high effect on the public – play a crucial role for the Transhumanist movement. A better understanding of their individual motivations and their role in technological development remains an ongoing challenge. Transhumanists aim at getting rid of the physical boundaries of humans, especially the ageing bodies and their permanent decline of functions. In the previous sections we have presented their vision and sketched a critique of their current content: we now want to come back to our perspective of vision assessment. Undoubtedly, the sort of vision Transhumanists present influences the shape of future technologies. This vision has an effect on the public perception of the technological innovation process, it attracts attention, it guides engineers and scientists, it fosters grants and funding and it drives the use of new and emerging technologies (Borup/Brown/Konrad/van Lente 2006). This is no different in the case of anti-ageing medicine. With certain resentment and followed by a warning, Olshanskhy and Leonard Hayflick wrote in the Scientific American in 2002:

"[...] the hawking of anti-aging "therapies" has taken a particularly troubling turn of late. Disturbingly large numbers of entrepreneurs are luring gullible and frequently desperate customers of all ages to "longevity" clinics, claiming a scientific basis for the anti-aging products they recommend and, often, sell. At the same time, the Internet has enabled those who seek lucre from supposed anti-ageing products to reach new consumers with ease." (Olshansky/Hayflick/Carnes 2002a: 92)

Some authors have seen this as a case to attribute responsibility in the development of new and emerging technologies (Simakova/Coenen 2013). As we observed, the sort of anti-ageing intervention Aubrey de Grey is envisioning lacks scientific evidence and his promises are as much as his arguments largely based on rhetoric sleights of hands: De Grey uses strong metaphors that rise negative associations to ageing. Nevertheless, one could agree that ageing is for many people accompanied by painful diseases and physical decline and, furthermore, that our society is indeed obliged to help people who permanently live in states of dependency. An ageing population poses challenges and bears duties for societies. In this respect, one has to be aware that Transhumanists currently focus only on technological means to face these challenges, which are less successful and less "scientific" than they are presented.

Furthermore, transhumanists disregard non-technological means of managing these problems. The vision of overcoming the ageing process narrows our view on the possible alternatives to deal with the problem of the ageing population and the lack of well-being of (ageing) individuals. Furthermore, this vision neglects the social and psychological state of elderly people, which needs to be taken into account in care practices. The habits that elderly people have established during their lives, for instance, are allegedly harder to be shaped in late life. This is a strong indicator of the amount of novelties that are bearable for them. Increasing the knowledge about this sort of psychological facts in aged populations could support the creation of appropriate care settings and treatments. Psychological and sociological research that follows this motivation may have fruitful results in understanding ageing, but this sort of knowledge is not in the foreground of the Transhumanist agenda. The social and economic conditions in which people live contribute substantially to their longevity and health. In general, people in western countries are healthier and grow older than people in the global South. People in the global South suffer from malnutrition and premature death due to diseases that are curable or at least treatable in western countries. These differences between socio-economic groups are not only to be found when we compare countries to each other, but also within countries these differences in life-expectancy can be verified for certain social classes and gender (Overall 2005). When we compare, for example, poorer people to richer people in the UK, we see that the average morbidity and mortality rate of the rich is clearly lower (Academy of Medical Sciences 2009). Another example is the population in Sardinia, which grows much older and sustains health until late life. Anthropological research has indicated that besides genetic and biological factors, the behavioural factors including life style, demographic behaviour, family support, and community characteristics may play an important role (Poulain/Pes/Salaris 2011).

These are complex phenomena, influential in the ageing process, which, as we have argued, Transhumanists do neglect in their argumentation. The pursuit to overcome physical barriers and preventing suffering are virtuous goals to aim at for policy makers, but they should be encountered in their complexity and balanced to the variety of technological and non-technological means to improve societal well-being. Besides Transhumanism there are other technological visions with other focuses that contest our values and shape our expectations. To develop reflective capacities to assess such visions and uncover their shortcomings could become increasingly useful.

NOTE

This article is largely based on the Master Thesis "Considerations on the Value of Ageing, Death and their Technological Mastering" of one of the authors, Martin Sand, written at the Institute of Technology Assessment and Systems Analysis in 2014.

REFERENCES

Academy of Medical Sciences (2009): Rejuvenating ageing research. A report by the Academy of Medical Sciences, London: Academy of Medical Sciences.

Achenbaum, A.W. (2005): "Ageing and Changing: International Historical Perspectives on Ageing". In: Johnson, M.L. (ed.), The Cambridge Handbook of Age and Ageing, Cambridge, New York: Cambridge University Press, pp. 21–29.

Bainbridge, R.S. (2013): "Transavatar". In: More, M./Vita-More, N. (eds.), The transhumanist reader. Classical and contemporary essays on the science, technology, and philosophy of the human future, Chichester, West Sussex, UK: Wiley-Blackwell, pp. 91–99.

Bengtson, V./Putney, N./Johnson, M.L. (2005): "The Problem of Theory in Gerontology Today". In: Johnson, M.L. (ed.), The Cambridge Handbook of Age and Ageing, Cambridge, New York: Cambridge University Press, pp. 3–20.

Blumenthal, H.T. (2003): "The Aging-Disease Dichotomy: True or False?". In: The Journals of Gerontology Series A: Biological Sciences and Medical Sciences 58, pp. 138–145.

Borup, M./Brown, N./Konrad, K./van Lente, H. (2006): "The sociology of expectations in science and technology". In: Technology Analysis & Strategic Management 18, pp. 285–298.

Bostrom, N. (2005): "The fable of the dragon tyrant". In: Journal of Medical Ethics 31, pp. 273–277.

Bostrom, N. (2013): "Why I want to be Posthuman when I grow up". In: More, M./Vita-More, N. (eds.), The transhumanist reader. Classical and contemporary essays on the science, technology, and philosophy of the human future, Chichester, West Sussex, UK: Wiley-Blackwell, pp. 28–53.

Broderick, D. (2001): The spike. How our lives are being transformed by rapidly advancing technologies, New York: Forge.

Carroll, L/Gardner, M./Tenniel, J.(2000): The annotated Alice. Alice's adventures in Wonderland & Through the looking glass, New York: Norton.

Coenen, C. (2006): "Der posthumanistische Technofuturismus in den Debatten über Nanotechnologie und Converging Technologies". In: Nordmann, A. (ed.), Nanotechnologien im Kontext. Philosophische, ethische und gesellschaftliche Perspektiven, Berlin: Akademische Verlagsgesellschaft, pp. 195–222.

Coenen, C (2009): "Zauberwort Konvergenz". In: Technikfolgenabschätzung – Theorie und Praxis 18, pp. 44–50.

Dierkes, M./Hoffmann, U./Marz, L. (1992): Leitbild und Technik. Zur Entstehung und Steuerung technischer Innovationen, Berlin: Edition Sigma.

Drexler, K.E. (1986): Engines of creation. The coming era of nanotechnology, New York: Anchor Books.

Ehni, H.-J. (2009): "Kann man sich Elina Makropoulos als glücklichen Menschen vorstellen? Ein Beitrag zur ethischen Debatte über den individuellen Wert eines längeren Lebens". In: Honnefelder, L./Sturma, D. (eds.), Jahrbuch für Wissenschaft und Ethik, Berlin, New York: Walter de Gruyter, pp. 47–70.

Ehni, H.-J. (2014): Ethik der Biogerontologie, Wiesbaden: Imprint: Springer VS.
Everts Mykytyn, C. (2010): "A history of the future: the emergence of contemporary anti-ageing medicine". In: Sociology of Health & Illness 32, pp. 181–196.
Fukuyama, F. (2003): Our posthuman future. Consequences of the biotechnology revolution, Princeton, N.J: Farrar, Straus and Giroux.
Grey, A. de (2003): "The foreseeability of real anti-aging medicine: focusing the debate". In: Experimental Gerontology 38, pp. 927–934.
Grey, A. de (2004): "The War on Aging". In: Immortality Institute (ed.), The scientific conquest of death. Essays on infinite lifespans, Buenos Aires: Libros en Red, pp. 29–45.
Grey, A. de (2005): "Resistance to debate on how to postpone ageing is delaying progress and costing lives". In: EMBO reports 6, pp. 49-53.
Grey, A. de/Rae, M./Burgermeister, P. (2010): Niemals alt! So lässt sich das Altern umkehren: Fortschritte der Verjüngungsforschung, Bielefeld: Transcript.
Gruman, G.J. (1966): A history of ideas about the prolongation of life. The evolution of prolongevity hypotheses to 1800, Philadelphia: American Philosophical Society.
Grunwald, A. (2009): "Vision Assessment Supporting the Governance of Knowledge – the Case of Futuristic Nanotechnology". In: Bechmann, G./Gorokhov, V./Stehr, N. (eds.), The social integration of science. Institutional and epistemological aspects of the transformation of knowledge in modern society, Berlin: Edition Sigma, pp. 147–168.
Grunwald, A. (2012): Technikzukünfte als Medium von Zukunftsdebatten und Technikgestaltung, Karlsruhe, Hannover: KIT Scientific Publishing; Technische Informationsbibliothek u. Universitätsbibliothek.
Grunwald, A. (2013): "Techno-visionary Sciences. Challenges to Policy Advice". In: Science, Technology & Innovation Studies 9, pp. 21–38.
Hayflick, L. (2000): "The Future of Ageing". In: Nature 408, pp. 267–269.
Heil, R. (2010): "Trans- und Posthumanismus. Eine Begriffsbestimmung". In: Hilt, A. (ed.), Endlichkeit, Medizin und Unsterblichkeit. Geschichte – Theorie – Ethik, Stuttgart: Steiner, pp. 127–149.
Huxley, J. (1931): "Biology and the Physical Environment of Man". In: What dare I think? The challenge of modern science to human action & belief. Including the Henry La Barre Jayne foundation lectures (Philadelphia) for 1931, New York, London: Harper & brothers, pp. 1–44.

Janich, P. (2009): Kein neues Menschenbild. Zur Sprache der Hirnforschung, Frankfurt: Suhrkamp.

Lenk, H. (ed.) (1994): Macht und Machbarkeit der Technik, Stuttgart: Reclam.

Lente, H.v. (1993): Promising technology. The dynamics of expectations in technological developments, Delft: Eburon.

McCay, C.M/Crowell, M.F. (1934): »Prolonging the Life Span«. In: The Scientific Monthly 39, pp. 405–414.

McCray, P. (2013): The visioneers. How a group of elite scientists pursued space colonies, nanotechnologies, and a limitless future, Princeton: Princeton University Press.

More, M. (2013a): "The philosophy of Transhumanism". In: More, M./ Vita-More, N. (eds.), The transhumanist reader. Classical and contemporary essays on the science, technology, and philosophy of the human future, Chichester, West Sussex, UK: Wiley-Blackwell, pp. 3–17.

More, M. (2013b): "The Proactionary Principle: Optimizing Technological Outcomes". In: More, M./Vita-More, N. (eds.), The transhumanist reader. Classical and contemporary essays on the science, technology, and philosophy of the human future, Chichester, West Sussex, UK: Wiley-Blackwell, pp. 258–267.

Murphy, D. (2015): "Concepts of Disease and Health", In: Zalta, E.N. (ed.), *The Stanford Encyclopedia of Philosophy* (Spring 2015 Edition), http://plato.stanford.edu/archives/spr2015/entries/health-disease/ (last access: 11.06.2015).

Nerlich, B. (2005): "From Nautilus to Nanobo(a)ts:. The Visual Construction of Nanoscience". In: AZojono – Journal of Nanotechnology Online 1, pp. 1–19.

Nipperdey, T. (1975): "Die Utopia des Thomas Morus und der Beginn der Neuzeit". In: Reformation, Revolution, Utopie. Studien zum 16. Jahrhundert, Göttingen: Vandenhoeck & Ruprecht, pp. 113–142.

Olshansky, S.J./Carnes, B.A. (2013): "Science Fact versus SENS Foreseeable". In: Gerontology 59, pp. 190–192.

Olshansky, S.J./Hayflick, L./Carnes, B.A. (2002a): "No Truth to the Fountain of Youth". In: Scientific American 286, pp. 92–95.

Olshansky, S.J./Hayflick, L./Carnes, B.A (2002b): "Position Statement on Human Aging". In: The Journals of Gerontology Series A: Biological Sciences and Medical Sciences 57, pp. 292-297.

Overall, C. (2005): Aging, death, and human longevity. A philosophical inquiry, Berkeley, London: University of California Press.

Poulain, M./Pes, G./Salaris, L. (2011): "A Population Where Men Live As Long As Women: Villagrande Strisaili, Sardinia". In: Journal of Aging Research, vol. 2011, pp. 1–10.

Roco, M.C./Bainbridge, W.S. (eds.) (2003): Converging technologies for improving human performance. Nanotechnology, biotechnology, information technology and cognitive science, Boston: Dordrecht.

Ropohl, G. (1991): Technologische Aufklärung. Beiträge zur Technikphilosophie, Frankfurt am Main: Suhrkamp.

Rose, M.R. (2004): "Biological Immortality". In: Immortality Institute (ed.), The scientific conquest of death. Essays on infinite lifespans, Buenos Aires: Libros en Red, pp. 17–28.

Rose, M.R. (2013): "Immortalist Fictions and Strategies". In: More, M./Vita-More, N. (eds.), The transhumanist reader. Classical and contemporary essays on the science, technology, and philosophy of the human future, Chichester, West Sussex, UK: Wiley-Blackwell, pp. 196–204.

Saage, R. (2006): "Konvergenztechnologische Zukunftsvisionen und der klassische Utopiediskurs". In: Nordmann, A. (ed.), Nanotechnologien im Kontext. Philosophische, ethische und gesellschaftliche Perspektiven, Berlin: Akademische Verlagsgesellschaft, pp. 179–194.

Saage, R. (2007): "Renaissance der Utopie". In: UTOPIE kreativ 201/202, pp. 605–617.

Sandberg, A. (2013): "Morphological Freedom". In: More, M./Vita-More, N. (eds.), The transhumanist reader. Classical and contemporary essays on the science, technology, and philosophy of the human future, Chichester, West Sussex, UK: Wiley-Blackwell, pp. 56–64.

Schade-Tholen, S./Franke, B. (1998): "Jungbrunnen und andere "Erneuerungsbäder" im 15. und 16. Jahrhundert". In: van Dülmen, R. (ed.), Erfindung des Menschen. Schöpfungsträume und Körperbilder 1500 – 2000; Wien, Köln, Weimar: Böhlau, pp. 197–218.

Schermer, M./Pinxten, W. (eds.) (2013): Ethics, health policy and (anti-)aging. Mixed blessings, Dordrecht, Heidelberg, London, New York: Springer.

SENS Foundation (2013): Annual Report, http://sens.org/sites/srf.org/files/reports/SENS%20Research%20Foundation%202012%20Annual%20Report.pdf (last access: 30.04.2015).

Sierra, F./Hadley, E./Suzman, R./Hodes, R.(2009): "Prospects for Life Span Extension". In: Annual Review of Medicine 60, pp. 457–469.

Simakova, E./Coenen, C. (2013): "Visions, Hype, and Expectations: a Place for Responsibility". In: Owen, R./Bessant, J.R./Heintz, M. (eds.), Responsible innovation. Managing the responsible emergence of science and innovation in society, Chichester: Wiley-Blackwell, pp. 241–266.

Sternberg, R./Grigorenko, E. (2005): "Intelligence and Wisdom". In: Johnson, M.L. (ed.), The Cambridge Handbook of Age and Ageing, Cambridge, New York: Cambridge University Press, pp. 209–215.

Various (2013): "Transhumanist Declaration". In: More, M./Vita-More, N. (eds.), The transhumanist reader. Classical and contemporary essays on the science, technology, and philosophy of the human future, Chichester, West Sussex, UK: Wiley-Blackwell, pp. 54–55.

Vincent, J. (2013): "The Anti-Aging Movement. Contemporary Cultures the Social Construction of Old Age". In: Schermer, M./Pinxten, W. (eds.), Ethics, health policy and (anti-) aging. Mixed blessings, Dordrecht, Heidelberg, London, New York: Springer, pp. 29–40.

Warner, H./Anderson, J./Austad, S./Bergamini, E./Bredesen, D./Butler, R./Carnes, B. A./Clark, B.F.C./Cristofalo, V./Faulkner, J./Guarente, L./Harrison, D.E./Kirkwood, T./Lithgow, G./Martin, G./Masoro, E./Melov, S./Miller, R.A./Olshansky, S.J./Partridge, L./Pereira-Smith, O./Perls, T./Richardson, A./Smith, J./Zglinicki, T.v./Wang, E./Wei, J.Y./Williams, T.F. (2005): "Science fact and the SENS agenda". In: EMBO reports 6, pp. 1006–1008.

Williams, B. (1993): "The Makropulos Case. Reflections on the Tedium of Immortality". In: Fischer, J.M. (ed.), The Metaphysics of death, Stanford: Stanford University Press, pp. 73–92.

Focusing on the Human: Interdisciplinary Reflections on Ageing and Technology

MARIA BEIMBORN, SELMA KADI, NINA KÖBERER, MARA MÜHLECK, MONE SPINDLER

Technology and society co-construct each other. Practices and perceptions of others and the self as well as the world outlook are highly mediated by technology; in turn, society – i.e. in the form of concepts and perceptions of age and ageing – is inscribed in technologies. Technologies can support specific ways of living together, but they can also be appropriated in varied ways. The choices concerning desired and avoided concepts of society that are made in technology development processes highlight the need for ethical reflections on technology.

The basic insight of that co-construction, derived from technology studies, not only sketches out a highly complex field for critical research, but also points to the ethical dimensions of technological development. Ethical questions, though, are only one reason for taking the human being as a starting point in the following reflections on ageing and technology. A limited understanding and consideration of the human factor in technological research risks to miss out some crucial aspects necessary for a rich understanding and a responsible shaping of a technologized ageing society.

How are older people and how is ageing imagined, conceptualized, inscribed and involved in current technological research? Looking into the field of technology development in Germany, technologies are explicitly developed for older people and/or the demographically changing society. Conceptualizations of age and ageing in this field are often shaped primarily by market research: we find these in the form of "human factors" and imagined "users" of these technological innovations. Recently technological development focuses on so-called interactive technologies, following the aim to bring forward technologies that are well adapted to the needs and preferences of the users. Thereby specific conditions of old age – quite often conceptualized as age-group-specific inabilities or disabilities – are taken into account in the technological design of the product. A recent trend is to increasingly involve the targeted age group in research and development since "user participation" promises two things at once: better adapted products and market success.

Our interdisciplinary perspective on ageing and technology aims to include a rich(er) understanding of older people as users of technologies and an ethical orientation within and for this field of research. We therefore take the older human being as a starting point.

We will first discuss concepts from the social sciences: These are a complex understanding of age and ageing developed in social gerontology, the analysis of the inscription of users, the appropriation of technologies by actual users developed in the user-script approach, and the analysis of interaction in socio-technical ensembles. The second part focuses on ethical questions linked to ageing and technology: We will reconsider the *conditio humana* from an ethical stance in order to introduce an ethically based concept of enabling/disabling technologies in a first step; in a second step, we propose that theories of the "good life" and concepts of "good care" and autonomy can provide orientation not only for technologies' implementation, use and evaluation, but also for research processes and research funding structures. In the third part, we ask how old people can participate in research. We discuss approaches such as the democratization of technology and meaningful participation as a basis for the analysis and evaluation of forms of involvement of older people in technological research.

In the conclusion we point out the implications for further research and discuss potentials for cooperation within the sciences as well as between the sciences and civil society in the field of ageing and technology.

1. PERSPECTIVES FROM THE SOCIAL SCIENCES ON HUMAN-TECHNOLOGY INTERACTION IN OLD AGE

A Differentiated Understanding of Age and Ageing

When taking human beings as the starting point of reflections on ageing and technology, one of the first questions is: What exactly characterizes older human beings? Images of ageing are quite influential in the development and use of technologies for older people. When a technology shall tackle "problems of ageing", there must be a notion involved of what age and ageing actually are.

In the 1990s, social gerontologists began to study images of ageing (Featherstone/Wernick 1995; BMFSFJ 2010: 35). They showed that stereotyped, reductionist, negative and therefore ageist images of ageing have until now been very popular in our culture. Countering these negative images with extra positive images of healthy, wealthy and always active seniors has, however, also proven to be a problematic strategy. Since both do not do justice to the complexity of age and ageing, it is still necessary to reflect on and differentiate images of ageing.

Against this backdrop many social gerontologists today share a certain discomfort about the way age and ageing are imagined in the context of technology (Oswald et al. 2008: 104). Especially in the field of assistive technologies, the ageing body is often seen as an increasingly malfunctioning machine, which itself needs technical support or surveillance. Furthermore, older people are frequently depicted as incapable of adapting to new technologies or even resistant to technology (Cutler 2005, Charness/Czaja 2005).

Here social gerontology can contribute to a differentiated understanding of age and age-ing. We argue that five essentials are important when taking older human beings as a starting point of reflections on ageing and technology: The fact that ageing is first a heterogeneous and second a multidimensional process; ageing is thirdly not linear decline but is located between strength and weaknesses; life courses (of older people) are fourthly deeply social and interconnected and fifth, that ageing is the subject of powerful discourses (Kruse/Wahl 2010: 77; Wahl/Heyl 2004: 41).

First, older people are not a homogeneous group, but a very diverse population. Their living conditions differ significantly according to, for example, income, education, gender, ethnicity, social networks, the housing situation, chronological age, the cohort a person belongs to and other categories of social stratification (Backes 1997; Heusinger et al. 2013). It is therefore difficult to generalize about "the needs of older people". Taking into account the heterogeneity of ageing rather reveals that usual patterns of social inequality continue across the life course: when technologies are assessed, the heterogeneity of ageing raises questions of inclusion and justice. Who does benefit from technological developments, who does not? Do older people have equal access to new technologies?

Ageing is, secondly, not only a biomedical bodily phenomenon. It has psychological, social, political, and economical as well as other dimensions

(Kruse/Wahl 2010: 336). The emphasis on the multidimensionality of ageing is based on a holistic view of human beings, according to which older people should not be reduced to a biomedical dimension of bodily decline. This raises anthropological questions for the assessment of technology: Which images of humanity and of older people underlie and foster a technology? It also points to the importance of including different disciplines in the development process, which can capture the multidimensionality of ageing.

Thirdly, ageing was long seen as a determined linear decline. Today, on the contrary, new medical approaches promise to prevent, stop or even reverse ageing. However, social gerontologists stress that ageing processes do not only go along with weaknesses, but also with strength and potentials (BMFSFJ 2006). Ageing is also not determined but – within limits – malleable. When technologies are assessed, this procedural view of ageing raises the question if a specific technology addresses only weaknesses or also strengths of ageing.

A fourth gerontological essential is that ageing should not be regarded as an isolated phase of an individual life. It rather is part of the life course, meaning a "sequence of age-linked transitions that are embedded in social institutions and history" (Bengtson et al. 2005: 493). The life course perspective reveals aspects of ageing which are often missing in popular images (ibid: 494): The individual life course is strongly interconnected with other people's life courses. Ageing processes can hardly be understood without their social and historical contexts and they need keen attention to biographical transitions. Older people do not "naturally" pass through "natural" stages of life but are active agents of their life courses.

From a critical gerontological perspective it is also important to, fifthly, take into account that our understanding of and dealing with ageing is shaped by powerful biomedical, economical and welfare state discourses: Since the beginning of modernity, a biomedicalization of ageing has taken place (Kaufman et al. 2004), which is problematic e.g. if the biomedical model leads to a situation where other dimensions of ageing are neglected. It is also well known that ageism is strongly linked to capitalism (Walker 2005), since potentially less "productive" older people become a problem only if productivity is the measure of all things. The strong political emphasis on encouraging older people to keep active, healthy and productive is part of a neoliberal welfare policy (van Dyk/Lessenich 2009), which tends to prefer individual responsibilities over societal concepts of solidarity. For the assessment of technology this raises the question whether a specific technology supports a problematic biomedicalization of ageing, a primarily economic view on ageing or emphasizes an activation of ageing without at the same time stressing the need for the improvement of the care system.

These gerontological essentials stress the need to think age and ageing as a much more complex process than negative or extra positive images of ageing suggest. Having sharpened our understanding of older human beings, we can now turn to the question of how they interrelate with technology.

Older People in Technology Development Processes: User Script Analyses

A social science perspective which analyses technology development as social process can highlight the contingent character of technical developments and power relations which shape them. An important strand of this work is the analysis of the co-construction of technology and society developed in the field of Science and Technology Studies (Hackett et al. 2007; Bijker et al. 2012). Together with the contribution in terms of a better understanding of technology in society, this can also be taken as a starting point for a combination of an integrated social science and ethics approach for studying ageing and technology. Such an integrated approach can use this analysis as a foundation for drafting future alternatives in terms of technology and the "good life" in old age (see section 2 of this chapter for a discussion of the "good life"). The analysis of technology development as social process begins with questions such as which technologies are developed, by whom, for whom and under which conditions.

This approach, which can be extended beyond the examination of the development of a particular technology and scrutinize the wider context in which it takes place, can be used to carve out existing trends in the broader field of "ageing and technology". An example for this is the analysis of discourses which underpin specific funding policies and decisions by private companies to invest in the development of specific products, such as a neoliberal discourse on individual self-responsibility combined with the promotion of active ageing (van Dyk 2009). Technologies are developed for diverse identified needs in old age (support for people with age-related illnesses, adaptations of information and communication technologies for older people, etc.) (BMBF/VDE Innovationspartnerschaft AAL 2011), but what opportunities are there to focus on human beings in analyses of technology development as social process?

There are multiple ways of focusing on older people in this approach. For example, older people and their experiences can be taken as key measure for the analysis and evaluation of technology development which aims for an improvement of life in old age. This can also be combined with an ethics and social science perspective which investigates trends in the field of "ageing and technology", comparing existing and envisioned technology against questions of the "good life in old age" (see section 2). Another approach is the inclusion of older people in research teams, for example as co-researchers (Walker 2007). We will now briefly discuss a third perspective for making older people central in the study of technology development as social process, namely to investigate inscribed and actual users of technologies.

This approach is based on Akrich's (1992, 1995) investigation of user scripts in technologies. She argued that designers merge various, explicit (e.g. market research) and implicit (e.g. personal experiences) user representations in the development process. Simultaneously, designers have to link technical options and potential markets for a technology. However, Akrich emphasizes that technologies emerge from the confrontation of user scripts with actual users, since the latter can appropriate technologies in unforeseen ways. For example, users can adapt technologies to solve different problems than those that the designers wanted to address. It is therefore important to study both inscribed and actual users to understand technology development. As Neven (2011) has demonstrated, the user script approach can be utilized to analyse dominant themes – such as the assumption that older people are always less competent technology users – in the inscription of older users in technology.

The advantage of the user script approach is that technology development is not studied in isolation from use; instead, a focus on user scripts suggests that, in order to understand technology, we have to equally study its appropriation. The inscription of users in technology goes beyond the identification of target user groups and considers that users are inscribed in technology in various ways, e.g. as individuals with relationships, specific capabilities and experiences and diverse characteristics (ergonomic, experiential, economic etc.). The analysis of inscribed and actual users can be taken as a base for a discussion of current trends in the field of "ageing and technology" and compared to a differentiated understanding of ageing from an integrated ethics and social sciences perspective.

Thinking Older People in Interaction with Technology

A social scientific perspective also brings the interaction of older people with technology into focus. Human-technology interaction is a tricky notion, particularly in interdisciplinary research. Since very different notions of that interaction exist, a careful explication of the concept is necessary:

In technology development, the term human-technology interaction usually refers to how humans operate technologies. Its notion of interaction was borrowed from computer sciences (Heesen 2014: 7); accordingly, the focus is put on the design of the technical interface, which – in this perspective – is the main site of interaction. A similar concept of human-technology interaction is often to be found in the development of AAL technologies for older people. Building on the assumption that older people are not used to inter-act with technology, the focus is put on adapting the technology to the reduced competences of older people. Simple or even invisible interfaces are one solution. Furthermore, increasingly intelligent and interactive technologies are developed, which are supposed to support individual older persons in everyday life by suggesting or by taking decisions (BMBF/VDE Innovationspartnerschaft AAL 2011: 12).

Here, a sociologically enriched notion of human-technology interaction allows for a more comprehensive understanding of that interplay. With regard to recent conceptual developments in technology studies for capturing the complex human-technology relations (Oudshoorn/Pinch 2007), we suggest to put four aspects into focus, namely: first, taking all actors into account, second, acknowledging the reciprocity, openness and contextuality of human-technology interaction, third, granting full agency to older users and fourth, understanding technology as a part of older people's lives.

First, human-technology interaction does not only take place between a single user and one technical device. The site of interaction rather is a socio-technical ensemble of multiple interrelated actors (Bijker 1995). Besides the individual user, his or her wider social environment is part of the interaction as well. Institutions can play an important role, too, which is hardly recognized so far (Oudshoorn/Pinch 2007: 556). Finally, technology does not only function but is itself regarded as a social actor in some recent sociological research (Rammert/Schulz-Schaeffer 2002; Latour 2007). This is because it shapes social practices and interactions and impacts on the perception of the self as well as the world. Human-technology interaction is therefore distributed between a complex network of humans, institutions and technologies. Understanding older people's interaction with technology therefore also includes how their relatives and friends, their neighbours and care givers, maybe also care homes or municipalities as well as the technology itself contribute to the interaction.

Understanding human-technology interaction sociologically means, secondly, that this interaction is no one way process but consists of mutually adaptive performances (Meister 2011): actors tend to anticipate each other's actions and adapt their actions accordingly. Thereby humans, societies and technology mutually constitute each other. Human-technology interaction therefore potentially changes all actors involved: The preferences of people can change as well as their role in a socio-technical ensemble. Technological artefacts can change, particularly the roles and values ascribed to them. The (power) relations between human, institutional and technical actors can in turn be altered. Human-technology interaction is therefore an open process and not determined, but conditioned by societal, political and cultural contexts.

Despite recent conceptual efforts to put the cultural appropriation of technology into focus (Hahn 2011), older people are still often not viewed as active actors in the human-technology interaction (Krummheuer 2010). The focus is rather put on their "unability" to adapt to new technologies (Charness/Czaja 2005: 662). With keen attention on different forms of interaction – verbal, bodily, intentional as well as unintentional interactions – we have to bring their diverse, potentially creative and subversive ways of technology appropriation into sight. It is therefore, thirdly, a matter of granting full agency to older people – without putting a case for an idealized notion of autonomy.

A fourth aspect is that particularly in the development of technologies for older people one finds the assumption that older people so far were isolated from technology, which now has to be introduced into their lives as a new factor. Philosophy and sociology of technology, however, have disproved this narrow understanding of technology (Hubig 2006): from their perspectives, human life is always mediated by technology, as we do not perceive the world independently from technology. For older people, human-technology interaction is a normal part of their lives, too (Pelizäus-Hoffmeister 2013, see also article in this volume). However, there are new technologies which – so far – hardly play a role in older people's lives.

These four aspects of a more differentiated, sociological understanding of human-technology interaction show that this complex interplay cannot be reduced to an individual person operating a technological device. Nor should the perspective be reduced to a triadic relation consisting of a technology, a care-taker and an old person. Rather the whole socio-technical ensemble has to come into sight. Human-technology interaction not least shapes relations between humans, between technologies, the relations of humans to themselves as well as between individuals and their social surroundings respectively the society. A critical analysis of technologies that are developed *for* older people should comprise all these different levels. Such perspective thereby also facilitates an in-depth analysis of power relations within human-technology ensembles i.e. how power is distributed within an interactive system, which role the technology design plays in ascribing certain roles and setting responsibilities, in creating meanings and confining spaces and scopes for agency.

The discussions around currently developed and introduced (interactive) technologies claim that through their potential to adapt to older users, these technologies meet the specific needs of people of old age and/or are appropriate means to meet the challenges of an ageing society. In the development of an approach for a human centred research we drew not only to social sciences but also to ethics. The human focused approach is both, a result and a quest of an intense transdisciplinary collaboration. Ethical perspectives can provide orientation for human centred technology development and use, as approaches of good life in old age and reflections on care urge to take the older person into account. When talking about technologies *for* older people, we refer to technologies that are able to enable good life and good care in old age. But how can the good life in old age be outlined? How to think technologies in ensembles of good care?

2. ETHICAL PERSPECTIVES ON GOOD LIFE IN OLD AGE

Taking the human being as a starting point of ethical reflections in the field of ageing and technology, we want to reconsider two things: firstly, the *conditio humana*, thus that what from a philosophical-anthropological perspective is defined to be the human nature or the basic human condition, and secondly, ethical approaches of the good life. By raising the basic question how we *want* to live as individuals and as societies we aim to ethically orientate the reflection on ageing and technology. Further we want to discuss the potentials of the good life in old age for the evaluation, development and practical use of (interactive) technologies.

The common narration of technology is that it is enabling: Innovation is often thought in technical terms and technical innovation stands, if not anymore for "progress" of societies, then still often for "improvement". In the following, we want to move away from the common understanding of "enabling technologies" characterized by aiming for the restoration of certain (dwindling) basic bodily functions or means that seem necessary for social participation, or focusing on an improvement of subjective life quality. Instead we pick up a conceptualization of "enabling/disabling technologies" developed in disability studies (Apelmo 2012), proposing two things: an anthropological-ethical grounding and an ethical orientation of the conception. The crucial questions we want to develop are: (How) Can technologies enable/disable the good life – for us as individuals and as a society? (How) Can technologies contribute to "good care" and a self-determined life in old age?

In the following, we argue that the stance of the good life offers a possible orientation and therefore reference point not only for an ethical impact assessment of technologies, but also for technological research and practical use of technologies. Looking to theories of the good life, we have to diagnose that they have long been blind to the fact of ageing. The good life, as it is framed in Western philosophical traditions, generally requires a healthy, rational, autonomous and highly reflected subject. What does that mean for (thinking) the good life in old age? Instead of developing a specific or reduced concept of the good life in old age(s), we need to rethink those theories from their essence. Our starting point is instead the vulnerable, dependent and related human being and the knowledge that ageing is by no means a unified, simple or linear process. Such an ethical approach towards the good life further needs to critically (re)consider questions of good care, concepts of autonomy and not least the role of technologies as they not only shape imaginations of ageing in society, but also practically enable/disable the good life. In the following, we want to give some orientation and continue the work that has to be done. Therefore we will first introduce Martha Nussbaum's capability approach and her ten dimensions of the good life, arguing that they could provide ethical orientation for the evaluation of technologies. In the second part, we consider questions of "good care" and introduce some (re) formulations of the concept of autonomy.

Enabling Technologies? Reconsidering the Approach of Capabilities

In Martha Nussbaum's influential capability approach, individual ethics and social ethics are strongly interlinked as her approach of a life in dignity is strongly connected to her concept of a good society (Nussbaum 2006, 2011). In the context of her reflections on human capacities in general, she formulates ten core capabilities which she further claims to be social entitlements.

Meeting these capabilities is not only a precondition for a dignified life for the individual but at the same time characterizes a good society.

We here refer to these core capabilities as they offer to outline a universal but culturally sensitive concept of (dignified) human life and good society. As the formulated capabilities are operationalizable, the list – that is considered as open – is a useful tool to ethically evaluate technologies and their use in terms of the good life. Instead of asking technical questions like if a certain technology restores or compensates dwindling physical, mental or communicative abilities, if it enables social participation – e.g. by improving mobility –, or if it improves subjective life quality, we can ask if it enables the development of human core capabilities. The approach thereby urges us to consider the specific societal and cultural conditions as they moderate capabilities e.g. are definitions of bodily integrity influenced by different religious regulations, societal discourses and cultural practices? Let us demonstrate this by operationalizing just a few of the ten dimensions for the evaluation of technologies, their development and use: Does certain (use of a) technology enable *bodily integrity*? Does it enable or rather disable *social affiliation*, e.g. good relations of care? Does a technology enable the older user's *practical reason* or, in other words, does it allow him/her to reflect on his/her own situation and enable him/her to develop an idea and plan of the own life? Does it enable the user to take *control over his or her environment*? Does a technology enable *play*? Answering such questions is no doubt a complex task and it requires careful consideration as it depends on social, institutional and cultural contexts, on practical use of technologies and not least on the cultural appropriations of a technology. To define the disabling and enabling potentials of technologies, ethics relies on a close collaboration with empirical social sciences. For the research of interactions in concrete ensembles of humans and technologies qualitative and especially ethnographic approaches seem to have most potential.

Questions of Good Care and Reconsidering Autonomy

To develop an ethically grounded concept of "good care" it is worth looking to the broad field of discourses on "care ethics". This predominantly feminist approach has not only strongly criticized the subject of classical philosophy and reformulated the *conditio humana,* the common conception of the human nature/condition, by claiming dependency, relatedness and vulnerability to be aspects of all human lives, but has also pointed to the political dimensions and ethical pitfalls of care (Brückner 2010). It is argued that one crucial problem of care – related to patriarchal societal order – is that although care is essential for human life, not only care-receiving is stigmatized, but also the different works of care-giving are dramatically devaluated. Another central problem is that care-relations are at risk of being shaped by paternalism, and ending up in domination and abuse. Dependence on care can endanger a life in dignity as it complicates self-respect and good affiliation. The need though is to conceptualize and experiment "good authority", which fosters and does not diminish self-respect. We have to develop an ethics of asymmetric (power) relations instead of generally considering such constellations as hindering for the good life and as disabling autonomy.

Empirical findings are thereby of crucial importance for ethical reflections. Lately, Andreas Kruse (Kruse, 2014) has published the results of two broad surveys, pointing out the importance of attention and social affiliation but also shared responsibility and caring for others as subjective dimensions of quality of life in old age. Considering on the one hand the enduring importance of *Weltgestaltung* and *Selbstgestaltung* in old age and on the other hand older people not only as care-receivers but also their wish to be care-givers asks us to redesign common ensembles of care and to rethink the political agenda of social participation as well as the orientation of care. Besides, the precarious conditions of institutionalized care, characterized by employee shortage, further economization and precarious work conditions demand to consider the human beings first – old people but also those who take care of them. To refer to terms of Hannah Arendt, it seems necessary to call for an orientation of care towards the human being as a person. Such an orientation towards the *who* instead of the *what* (Arendt 1998) asks if technology enables relations of care in which the individual is able to express him or herself and to find acceptance, as the preconditions for a life in dignity are mutual respect, social recognition and the ability to shape one's own life and participate meaningfully in society.

Starting from the vulnerable, related self, questions of good care also lead us to scrutinize our concept of autonomy. Focusing on age and ageing, we are urged to consider situations of dependency and conditions characterized by limited or unsteady mental and bodily capacities. "Classical" conceptions of autonomy, presupposing the freedom of will and the ability to reflect upon one's own person, thoughts and decisions appear somehow quixotic. Instead, we want to point to two approaches that basically reformulate autonomy as relational and gradual. Looking toward life in old and very old age helps to recognize two things: firstly that it is often rather a collective of people – in the named case the old person, relatives, doctors and other care-receivers – who take both essential and daily decisions and secondly that cognitive conditions are generally plural, dynamic and in certain phases of life (potentially) limited. Feminist scholars have argued that autonomy in practice is always relational and that the self is constituted in and through social relations (Holstein et al. 2011; Mackenzie/Stoljar 2000). In different fields – e.g. regarding medical decisions – autonomy has been reconceptualized as relational and gradual (Kipke/Rothhaar 2009; Downie/Llewellyn 2012).

Taking an approach to the good life that is based on rich concepts of ageing human beings as orientation for technology, we can formulate concrete questions towards technologies and the socio-technical ensembles and their funding and development practices: Does a certain (use of a) technology enable or rather disable us to live a good life – as individuals and as a society? Does it enable or rather disable relations and socio-technical ensembles of good care, characterized by a life in dignity with mutual respect, social recognition and participation? Does a technology support or is it rather hindering for a self-determined life in old age?

3. PARTICIPATORY PERSPECTIVES ON AGEING AND TECHNOLOGY

A third perspective of focussing on the human in the field of ageing and technology lies across and beyond the traditional academic disciplines: it consists in the participation of older people in the research on and in the development of technologies. Funding agencies, technicians as well as seniors call for a better integration of users in the development of technology for older adults. And indeed, older people are increasingly involved in development processes, for instance in the evaluation of products, in selected decisions or via empirical surveys on users' preferences (e.g. Glende et al. 2011). The rationale is that user integration leads to technical devices which are better adapted to the users and therefore to market success.

We share the call for more participatory designs in the field of ageing and technology. However, in order to sharpen our focus on the human, we have to rethink participation as currently applied in the field in three regards: Why is participation important? Who should participate? And how should participation be implemented?

First, we have to rethink the aims of participation. Why is it important that (older) people participate? The answer to this question reveals normative presuppositions which are crucial for the orientation of participation processes and the way participants are addressed: as users, as consumers, as experts or as citizens. Participation approaches developed in the context of direct democracy and action research (Bergold/Thomas 2010) show that participation can serve more than the goals of technology acceptance and market success. It is about the involvement of citizens; participation from this perspective is a civil right of people to shape areas and practices which are particularly relevant for their lives. The involvement of stakeholders is a political act to practice and to enforce the social participation, particularly of social groups that dispose of weak interest representation. In this sense, participation is not primarily about the optimisation of products. It is about democratization – in our case of the research on and the development of technology for older people. It aims at social change (Götsch et al. 2012: 43) and at fostering empowerment (Bergold/Thomas 2010: 49).

What does that mean with regard to involving older people? Since the 1980s, senior organisations along with social gerontologists have called for more participation in different fields. In the past decades, opportunities for older people to participate have indeed expanded (Barnes 1999), particularly at municipal level e.g. in the planning of health care services or housing policy but also in gerontological research. One reason for that expansion is that the relation between individual and state has changed in a way that older people are increasingly perceived as consumers of welfare services (Walker 2007). However, the original point of reference in seniors' and gerontologists' calls for more participation is also the ideal of a deliberative democracy and emancipation (Heslop 2002; Barnes 1999).

Furthermore, there are research ethical reasons why older people should be involved (Walker 2007: 482). The general right of people to be consulted about research that is conducted on them is particularly important for older adults, due to their experiences of ageism. Researchers, politicians and developers have to be careful not to reproduce such forms of discrimination and exclusion. Especially when research aims at improving the quality of life of older people, it is a matter of justice to consult them. Alan Walker takes assistive technologies as an example for "the inadequacy of attempts to involve older people in identifying needs and appropriate solutions." (ibid:482).

Who should participate? A second highly political question relates to the criteria according to which participants are chosen (Wright 2011) which is in turn linked to the aims of participation processes. If the aim is to take the diversity of older people's lives into account, it is necessary to recruit participants who can represent this diversity. To achieve this, it is useful to ask which groups of older people are affected by the development of a technology in different ways. If the aim is a more inclusive debate informed by the concept of a deliberative democracy, it is necessary to critically think about the conditions for participation, and to create a space which is receptive to various ways of engaging in a debate, rather than privileging only one (Barnes 1999). If the focus of participation is empowerment of older people, the acquisition of competences, such as training as a researcher or co-researcher (Walker 2007), can be an integral part of participatory designs. As Barnes (1999) suggests, it might also be useful to rethink the focus on participatory approaches which are based on institutionalized representation of older people. Many potential participants will bring with them experiences of other forms of involvement in decision making e.g. on a local political level. She argues that the formation of a collective identity of older people might be a result rather than a precondition for participation processes.

The third question is how participation should be implemented. Since participation has become a catchword in different contexts, there is a growing opinion among advocates of participation calling for full, for systematic or for meaningful participation. What makes participation in practise meaningful? Many of the arguments put forward in this context go back to questions of power. Usually the resources and the power to define and to decide is distributed unequally between those who participate (e.g. older people) and those who grant them the opportunity to get involved (e.g. engineers). A constant reflection on the distribution of power in the transdisciplinary team is essential (Götsch et al. 2012: 34; Bergold/Thomas 2010: 340, see also the contribution in this volume (Domínguez-Rué and Nierling 2015). We also have to ask how much influence is actually granted to the participants. When implementing participation, several aspects are crucial to deal responsibly with the asymmetric power relations characteristic for the involvement of stakeholders:

The participants should not be regarded and addressed as lay persons, but as experts in their lifeworld (*Lebenswelt*). Their competences and experiences should be acknowledged equally. This helps participants/experts to contribute and exchange their visions, skills and expertise more at eye level, which requires equal access to information and keen attention on the integration of limited or non-verbal forms of communication. It is also important to disclose the (different) expectations in the transdisciplinary team throughout the participation process. The gains of the project should not be one-sided but a win-win situation.

Another crucial point is the financial side of participation. If older people are involved in technology development, they usually do not receive a salary as their counterparts in the transdisciplinary team do. As salary is the common and socially most accepted form of acknowledgement, everyone who is working on the process of development – participatory personnel as well as classical developers – should be remunerated (Bergold/Thomas 2012: 37).

The question about the moment in a development process in which participation should take place is often also a question of power. When people are involved from the beginning through the whole process of technology development, they have more opportunities to negotiate even aims, research questions and structures which are decisive for the development process. If participation takes place for the evaluation of an already developed technical device, there are less opportunities to influence the process.

The accomplishment and the dealing with results is another important point. Participation can only be effective if research and development are adaptive processes which are open to really include results of participation in the final outcome. A difficult question here is how to negotiate conflicting ideas and opinions. An issue that is also largely unresolved is how different forms of knowledge – particularly experience-based knowledge of people affected and academic knowledge of researchers – should be integrated during participation processes. However, it is crucial to make transparent how the final results were accomplished and to ensure equal access to them.

In the practical implementation of participation, there is often the need to balance the far-reaching claims of a meaningful involvement with practical limits. It is important to ask under which conditions it is tolerable or necessary to limit participation e.g. by restraining participation to selected decisions during the development process (e.g. Lindsay et. al. 2012).

Why should who participate how? As transparency is a precondition for good participation (Bergold/Thomas 2010: 340); it is important to consider these basic questions before the starting point of policy making, research or technology development. They should be negotiated and communicated within the transdisciplinary team but also externally in order to reveal what exactly is meant by letting older people participate.

4. Doing Transdisciplinary Research on Technology in an Ageing Society

We have combined different research approaches, theories and methods to outline the potentials of cooperation of different disciplines and civil society that leads to both a rich and critical understanding of ageing and technology and an ethically as well as democratically orientated research and development of technology. Our point of reference are societies that are facing an unprecedented demographic change and that have decided to explicitly invest in technological innovations to manage, shape and control this societal change.

Integrating these various perspectives and actors into a framework of inter- or transdisciplinary research is costly and experimental. It is costly not only in terms of time and money, but also in that it requires a critical reflection of (disciplinary) concepts and rationalities of all actors involved. It is experimental in the sense that new research designs and new forms of participation in decision making – of other disciplines and/or civil society – will change accustomed paths of technological development and common distributions of power and authority. Sciences and humanities are asked to collaborate closely in the same way that science and civil society need to learn how different forms of knowledge can come together in a productive way. The success of such kinds of research projects relies on interdisciplinary competences, interdisciplinary spaces and the establishment of transdisciplinary structures and competences as well. This requires willingness to engage with other perspectives.

We started our reflections on age and ageing by pointing to the need for a more complex understanding of age and ageing in technology research and technological development. By outlining conceptions and research approaches that focus on older people as both imagined and real users of technologies, we underlined the need for a critical reflection on user scripts (within technological development processes) but also called for a broader research on the actual use and cultural appropriations of (new) technologies, considering that the impact of technologies highly depends on contexts and user practices. As these practices are (potentially) diverse and (can) change over time, ethical evaluations that only focus on the development process and the initial implication phase do not close the case. We urged to understand older people as active parts of socio-technical ensembles and therefore propose to consequently integrate a socio-scientific empirical approach, as it promises to be productive for the evaluation and design of interactive technologies and user specific interfaces.

The second part showed the starting point for our reflection and analysis: It is a concept of the good life that *includes* old age instead of isolating it as an (exceptional) stage of life. Our anthropological core statement is an understanding of human beings as fundamentally vulnerable and dependent on others in every phase of life – not just in childhood or old age. Enabling technologies thus relates to reflections of the good life and of good care, thus also rethinking classical concepts of autonomy.

As we see the need for broader ethical reflection in the context of technological development for older people, interdisciplinary approaches have to be further established. Ethical theories of the good life can thereby inform and advise democratic discussions and political processes but never substitute them.

In the third part we analyzed questions of participation of older people in technological development, reflecting on the potentials and pitfalls of established or intended practices. Which approach and method to follow is highly dependent on aims and resources. What is necessary, however, is a systematic reflection of the goals of participation, the transparency of these goals and processes and the reflection of research conditions that allow a meaningful involvement of older people. The current increase in participatory designs has great potentials but still needs conceptual work. Only then can technologies do justice to the wants and needs of older users through a better understanding of ageing which can derive from close contact on the one side and the democratization of technology on the other.

Society and technology co-construct each other. But (how) can technological research contribute to good life and good society? To develop a conception of the good life that embraces questions of good care, neither the horror scenario depicting a wave of incompetent old people needing (too) costly care, nor a reconfiguration of old people as active, healthy, and useful – and therefore not needing costly care – seems helpful. Instead we propose starting from an anthropological core statement: human beings are vulnerable and dependent on others in every stage of life. Old age is then not an exceptional stage but is rather to be understood as a radicalization of a human fundamental situation. Taking a glimpse at current innovations, we come across a lot of promises that suggest that technologies are enabling. They enable mobility, social participation, long life at home, security, health etc. But do they enable the older person and society to lead the good life?

Do they support and create good care? Especially so-called interactive technical innovations suggest that care relations will change through the use of new technologies. Created and supported by interactive technologies, highly engineered and cross-linked care ensembles are about to emerge. These "care platforms" link different human and technical care receivers, and collect, save, and organize information with different information technologies. Emotional intelligent technologies suggest that soon care will also embrace new forms of social relatedness between technologies and old people.

Facing the common imaginations of age and ageing and the accelerated technological development that allegedly is aimed at meeting the needs of the elderly and an ageing society, inter- and transdisciplinary approaches call those who research and develop technology to reflect on the ethical dimensions of their own action and acknowledge their responsibilities for shaping and organizing the future society – one in which vulnerability, dependence and relatedness will neither disappear nor be technically "solved", but rather needs to be acknowledged as the fundamental human condition.

REFERENCES

Akrich, M. (1992): "The De-Scription of Technical Objects". In: Bijker, W.E./Law, J. (eds.), Shaping Technology/Building Society. Studies in Sociotechnical Change, Cambridge, Massachusetts: MIT Press, pp. 205–224.

Akrich, M. (1995): "User Representations: Practices, Methods and Sociology". In: Rip, A./Misa, T.J/Schot, J. (eds.), Managing Technology in Society. The Approach of Constructive Technology Assessment, London, New York: Pinter Publishers; pp. 167–184.

Apelmo, E. (2012): "Falling in love with a wheelchair: enabling/disabling technologies". In: Sport and Society 15/3, pp. 399–408.

Arendt, H. (1998): The human condition, Chicago: University of Chicago Press.

Backes, G. (1997):" Lebenslage als Konzept zur Sozialstrukturanalyse". In: Zeitschrift für Sozialreform 43/9, pp. 704–727.

Barnes, M. (1999): "The same old process? Older people, participation and deliberation". In: Ageing and Society 25/2, pp. 245–259.

Bengtson, V.L./Elder, G.H./Putney, N. (2005): "The Lifecourse Perspective on Ageing: Linked Lives,Timing and History". In: Johnson, M.L (ed.),

The Cambridge Handbook of Age and Ageing, Cambridge, New York: Cambridge University Press, pp. 493–501.

Bergold, J./Thomas, S. (2010): "Partizipative Forschung". In: Mey, G. (ed.), Handbuch Qualitative Forschung in der Psychologie, Wiesbaden: VS Verlag für Sozialwissenschaften, pp. 333–444.

Bijker, W.E. (1995): Of Bicycles, Bakelites, and Bulbs. Toward a Theory of Sociotechnical Change, Cambridge, Massachusetts: MIT Press.

Bijker, W.E./Hughes, T.P./Pinch, T.J. (eds.) (2012 [1987]): The Social Construction of Technological Systems. New Directions in the Sociology and History of Technology, Cambridge, Massachusetts: MIT Press.

BMBF/VDE Innovationspartnerschaft AAL (ed.) (2011): Ambient Assisted Living (AAL). Komponenten, Projekte, Services, Band 3, Berlin, Offenbach: VDE.

BMFSFJ (Bundesministerium für Familie, Senioren Frauen und Jugend) (ed.) (2006): Fünfter Bericht zur Lage der älteren Generation in der Bundesrepublik Deutschland. Potenziale des Alters in Wirtschaft und Gesellschaft, Berlin: BMFSFJ.

BMFSFJ (Bundesministerium für Familie, Senioren Frauen und Jugend) (ed.) (2010): Sechster Bericht zur Lage der älteren Generation in der Bundesrepublik Deutschland. Potenziale des Alters in Wirtschaft und Gesellschaft, Berlin: BMFSFJ.

Brückner, M. (2010): "Entwicklungen der Care-Debatte – Wurzeln und Begrifflichkeiten". In: Apitzsch, U. (ed.), Care und Migration. Die Ent-Sorgung menschlicher Reproduktionsarbeit entlang von Geschlechter- und Armutsgrenzen, Opladen, Farmington Hills: Budrich, pp. 43–58.

Charness, N./Czaja, S.J. (2005): "Adaption to New Technologies". In: Johnson, M.L. (ed.), The Cambridge Handbook of Age and Ageing, Cambridge, New York: Cambridge University Press, pp. 662–669.

Cutler, S.J. (2005): "Ageism and Technologie." In: Generations 29/3, pp. 67–72.

Downie, J.G./Llewellyn, J.J. (eds.) (2012): Being relational. Reflections on Relational Theory and Health Law, Vancouver, British Columbia: UBC Press.

Featherstone, M./Wernick, A.(eds.) (1995): Images of Aging. Cultural Representations of Later Life, London: Routledge.

Glende, S./Nedopil, C./Podtschaske, B./Stahl, M./Friesdorf, W. (2011): Erfolgreiche AAL-Lösungen durch Nutzerintegration. Ergebnisse der Studie "Nutzerabhängige Innovationsbarrieren im Bereich Altersgerechter Assistenzsysteme", Berlin. http://www.youse.de/documents/Kompetenzen/Publikationen/YOUSE_AwB_2012_Erfolgreiche_AAL-Lsungen. PDF (last access: 04.08.2014).

Götsch, M./Klinger, S./Thiesen, A. (2012): ",Stars in der Manege?' Demokratietheoretische Überlegungen zur Dynamik partizipativer Forschung". In: Forum Qualitative Sozialforschung/Forum: Qualitative Social Research 13/1, Art. 4, http://www.qualitative-research.net/index. php/fqs/article/view/1780/3296 (last access: 06.08.14).

Hackett, E.J./Amsterdamska, O./Lynch, M.E./Wajcman, J. (eds.) (2007): The New Handbook of Science and Technology Studies, Cambridge, Massachusetts: MIT Press.

Hahn, H.P. (2011): "Antinomien kultureller Aneignung: Einführung". In: Zeitschrift für Ethnologie 136/1, pp. 11–26.

Heesen, J. (2014): "Mensch und Technik. Ethische Aspekte einer Handlungspartnerschaft zwischen Personen und Robotern". In: Hilgendorf, E. (ed.), Robotik im Kontext von Recht und Moral, Baden-Baden: Nomos, pp. 190–205.

Heslop, M. (2002): Participatory Research with Older People. A Sourcebook, London: HelpAge International, http://www.globalaging.org/elderrights/world/sourcebook.pdf (last access: 17.07.14).

Heusinger, J./Kammerer, K./Wolter, B. (2013): Alte Menschen. Expertise zur Lebenslage von Menschen im Alter zwischen 65 und 80 Jahren, Köln: BZgA.

Holstein, M./Parks, J.A./Waymack, M.H. (2011): Ethics, Aging, and Society. The Critical Turn, New York: Springer.

Hubig, C. (2006): Technikphilosophie als Reflexion der Medialität. Grundlinien einer dialektischen Philosophie der Technik Band 1: Technikphilosophie als Reflexion der Medialität, Bielefeld: Transcript.

Kaufman, S.R./Shim, J.K./Russ, A.J. (2004): "Revisiting the Biomedicalization of Aging: Clinical Trends and Ethical Challenges". In: The Gerontologist 44/6, pp. 731–738.

Kipke, R./Rothhaar, M. (2009): "Die Patientenverfügung als Ersatzinstrument. Differenzierung von Autonomiegraden als Grundlage für einen angemessenen Umgang mit Patientenverfügungen". In:. Frewer, A/Fahr, U./Rascher, W. (eds.), Patientenverfügung und Ethik. Beiträge zur guten klinischen Praxis. Jahrbuch Ethik in der Klinik 2, Würzburg: Königshausen & Neumann, pp. 61–75.

Krummheuer, A. (2010): Interaktion mit virtuellen Agenten? Zur Aneignung eines ungewohnten Artefakts, Stuttgart: Lucius & Lucius.

Kruse, A. (2014): Der Ältesten Rat. Generali Hochaltrigenstudie. Teilhabe im hohen Alter. Eine Erhebung des Instituts für Gerontologie der Universität Heidelberg mit Unterstützung des Generali Zukunftsfonds, Köln, http://zukunftsfonds.generali-deutschland.de/online/portal/gdinternet/zukunftsfonds/content/314342/1010874 (last access: 20.08.2015)

Kruse, A./Wahl, H.-W. (2010): Zukunft Altern. Individuelle und gesellschaftliche Weichenstellungen, Heidelberg: Spektrum.

Latour, B. (2007): Eine neue Soziologie für eine neue Gesellschaft. Einführung in die Akteur-Netzwerk-Theorie, Frankfurt am Main: Suhrkamp.

Lindsay, S./Jackson, D./Ladha, C./Ladha, K./Brittain, K./Olivier, P. (2012): Empathy, participatory design and people with dementia. CHI ` 12 May 5-10 Austin, Texas. New York: ACM.

Mackenzie, C./Stoljar, N. (eds.) (2000): Relational Autonomy. Feminist Perspectives on Automony, Agency, and the Social Self, New York: Oxford University Press.

Meister, M. (2011): "Mensch-Technik-Interaktivität mit Servicerobotern. Ansatzpunkte für eine techniksoziologisch informierte TA der Robotik". In: Technikfolgenabschätzung – Theorie und Praxis 20/1, pp. 46–52.

Neven, L. (2011): Representations of the old and ageing in the design of the new and emerging: assessing the design of ambient intelligence technologies for older people, Enschede: University of Twente.

Nussbaum, M.C. (2006): Frontiers of justice. Disability, Nationality, Species Membership, Cambridge, Massachusetts: The Belknap Press, Harvard University Press.

Nussbaum, M.C. (2011): Creating Capabilities. The Human Development Approach, Cambridge, Massachusetts: Belknap Press, Harvard University Press.

Oswald, F./Claßen, K./Wahl, H.-W. (2009): "Die Rolle von Technik bei kognitiven Einbußen im Alter". In: Landesstiftung Baden-Württemberg (ed.), Training bei Demenz. Schriftenreihe der Landesstiftung Baden-Württemberg 42, Stuttgart: Landesstiftung Baden-Württemberg, pp. 104–143.

Oudshoorn, N./Pinch, T. (2007): "User-Technology Relations: Some recent Developments." In:. Hackett, E.J/Amsterdamska, O./Lynch, M.E./Wajcman, J. (eds.), The New Handbook of Science and Technology Studies, Cambridge, Massachusetts: MIT Press, pp. 541–566.

Pelizäus-Hoffmeister, H. (2013): Zur Bedeutung von Technik im Alltag Älterer. Theorie und Empirie aus soziologischer Perspektive, Wiesbaden: Springer VS.

Rammert, W./Schulz-Schaeffer, I. (2002): "Technik und Handeln. Wenn soziales Handeln sich auf menschliches Verhalten und technische Abläufe verteilt." In: Rammert, W./Schulz-Schaeffer, I. (eds.), Können Maschinen handeln? Soziologische Beiträge zum Verhältnis von Mensch und Technik, Frankfurt am Main, New York: Campus, pp. 11–64.

van Dyk, S. (2009): "'Junge Alte' im Spannungsfeld von liberaler Aktivierung, Ageism und Anti-Aging-Strategien." In: van Dyk, S./Lessenich, S. (eds.), Die jungen Alten. Analysen einer neuen Sozialfigur, Frankfurt am Main, New York: Campus-Verlag, pp. 316–339.

van Dyk, S./Lessenich, S. (2009): "'Junge Alte': Vom Aufstieg und Wandel einer Sozialfigur." In: van Dyk, S./Lessenich, S. (eds.), Die jungen Alten. Analysen einer neuen Sozialfigur, Frankfurt am Main, New York: Campus, pp. 11–48.

Wahl, H.-W./Heyl, V. (2004): Gerontologie. Einführung und Geschichte, Stuttgart: W. Kohlhammer.

Walker, A. (2005): "25th volume celebration paper: Towards an international political economy of ageing". In: Ageing and Society 25/06, pp. 815–839.

Walker, A. (2007): "Why involve older people in research?" In: Age and Ageing 36/5, pp. 481–483.

Wright, D. (2011): "A framework for the ethical impact assessment of information technology". In: Ethics and Information Technology 13/3, pp. 199–226.'

List of Authors

Abdullah, Boushra is a 3rd year undergraduate student in Community Rehabilitation in the Cumming School of Medicine at the University of Calgary, Canada. Bushra's interests include aging well and education for sustainable development; bushraa1994@hotmail.com.

Adam, Frederic is Professor at the Department of Accounting, Finance and Information Systems in University College Cork, Ireland. His research interests include decision making and decision support, enterprise wide systems, and Health Information Systems and the impact of ICT on the delivery of healthcare; fadam@afis.ucc.ie.

Beimborn, Maria is research assistant at the International Centre for Ethics in the Sciences and Humanities (IZEW) at Eberhard Karls University Tübingen, Germany. Her research interests are in technology ethics and the ethnography of state and citizenship; Maria.beimborn@lmu.de.

Biniok, Peter, Dr. is research assistant in the Faculty of Health, Safety and Society at Furtwangen University, Germany. His main research interests are science and technology studies, innovation studies, man-machine-interaction, and technology at work; bip@hs-furtwangen.de.

Cook, Glenda, is Professor in the Faculty of Nursing at Northumbria University, UK. Her research generates knowledge concerning gerotechnology, older peoples' involvement in service planning and policy making, and the experience of living in residential and nursing homes; glenda.cook@northumbria.ac.uk.

Decker, Michael, is Professor at the Institute for Technology Assessment and System Analysis - Karlsruhe Institute of Technology, Germany. His areas of research include theory and methodology of TA, TA of new and emerging sciences and technologies and epistemology of inter- and transdisciplinary knowledge; michael.decker@kit.edu.

Domínguez-Rué, Emma, PhD, is Associate Professor in the Department of English at the University of Lleida, Catalunya (Spain). Her research interests include ageing and gender in literature and culture, contemporary crime fiction, feminism, psychoanalysis and mental disorders; edominguez@dal.udl.cat

Endter, Cordula, MA, works at the Institute of Cultural Anthropology, University of Hamburg, Germany. Her research focuses on technology and age, but also on other topics concerning age and aging like work, gender and mobility; cordulaendter@googlemail.com.

Fernández-Ardèvol, Mireia, PhD, is senior researcher at the IN3, Internet Interdisciplinary Institute – Open University of Catalonia. Mobile communication is one of her main areas of study since 2003, with a combined sociological and economic focus, both in developed and in developing countries; mfernandezar@uoc.edu.

Fitzgerald, Ciara, Dr., is Lecturer in the Department of Accounting, Finance and Information Systems at University College Cork, Ireland. Her research is focused on innovation, entrepreneurship and strategy, with a particular emphasis on the actors and activities involved in the implementation of innovation policy; cfitzgerald@ucc.ie.

Graner-Ray, Sheri, is CEO/Founder of Zombie Cat Studios, Inc., USA. Her research interests focus on gender and games; sgranerray@gmail.com.

Guihen, Barry is a doctoral researcher at the Centre for Professional Ethics in the University of Central Lancashire, UK. He has worked with the Centre for Science, Society and Citizenship, on ethical and societal issues with emerging technologies; bguihen@gmail.com.

Jongsma, Karin is a Phd Candidate at the Department of Medical Ethics and Philosophy of Medicine at the Erasmus Medical Centre, Rotterdam (Netherlands). During the course of her Phd research on dementia she worked as an advisor for the National Council for Public Health; k.jongsma@erasmusmc.nl.

Kadi, Selma, Dr., is postdoctoral researcher at the Institute for Ethics and History of Medicine at Eberhard Karls University Tübingen, Germnay. Her research interests centre in the fields of sociology of technology, sociology of ageing, feminist technology studies and intersectionality research; selma.kadi@uni-tuebingen.de.

Kamphof, Ike, PhD, is assistant professor in the Faculty of Arts and Social Sciences at Maastricht University (Netherlands). Her research, which combines phenomenology with ethnography, focuses on identity, embodiment and changing values in technologically mediated networks of care; i.kamphof@maastrichtuniversity.nl.

Klein, Barbara, is Professor for organization and management in social work at the Faculty of Social Work and Health at the Frankfurt University of Applied Sciences, Germany. Barbara leads a research group on independent living using assistive technologies for information, education, counseling and research; bklein@fb4.fra-uas.de.

Köberer, Nina, Dr., is research assistant at the International Centre for Ethics in the Sciences and Humanities (IZEW) at Eberhard Karls University Tübingen, Germany. Her research interests are in the field of media ethics, technical ethics, and narrative ethics; n.koeberer@gmx.de.

Krings, Bettina-Johanna, Dr., works at the Institute for Technology Assessment and System Analysis - Karlsruhe Institute of Technology, Germany. Her areas of research are technical innovations and their impact on work structures, theory and methods of TA, and theory and practice of gender research; bettina-johanna.krings@kit.edu.

Mantovani, Eugenio, LLM, is a doctoral researcher in the interdisciplinary Research Group on Law Science Technology & Society at the Vrije Universiteit Brussel, Belgium. Eugenio's research interests spawn across the area of regulation of technology with a specific focus on the domains of ageing and health; eugenio.mantovani@vub.ac.be.

Marston, Hannah R., Dr., is a research associate in the Centre for Research in Computing at the Open University, UK. Hannah's research interest include barriers and enablers to technology use/adoption, design, usability of mHealth apps and technologies for health and well-being; marstonhannah@hotmail.com.

Menke, Iris, is research assistant in the Faculty of Health, Safety and Society at Furtwangen University. Her research interests focus on urban anthropology and biographic research; meni@hs-furtwangen.de.

Moyle, Wendy is Professor and Director of the Centre for Health Practice Innovation in the Griffith Health Institute at Griffith University, Brisbane, Australia. The focus of her work is on dementia, assistive technologies, social robots and complementary and alternative medicine interventions; w.moyle@griffith.edu.au.

Mühleck, Mara is research assistant at the International Centre for Ethics in the Sciences and Humanities (IZEW) at Eberhard Karls University Tübingen. Her research interests are in qualitative methodology, especially ethnography, interaction theory and sociology of technology; mara.muehleck@izew.uni-tuebingen.de.

Nierling, Linda, Dr., is a researcher at the Institute for Technology Assessment and Systems Analysis (ITAS), Karlsruhe Institute of Technology (KIT), Karlsruhe (Germany). Her research interests include work and technology, conceptions and methods of technology assessment, assistive technologies, sustainability; nierling@kit.edu

Pelizäus-Hoffmeister, Helga, is Professor in the Faculty of Political and Social Science at the University of the German Federal Armed Forces in Munich. Her areas of research include demographic change, elderly and technology, qualitative research and biography research; Helga.pelizaeus-hoffmeister@unibw.de.

Sand, Martin, is a PhD Candidate at the Institute of Technology Assessment and Systems Analysis - Karlsruhe Institute of Technology, Germany. In his current research he focuses on problems of responsibility in the early, visionary stages of technological development; martin.sand@kit.edu.

Selke, Stefan, is Professor for sociology and social change in the Faculty of Health, Safety and Society at Furtwangen University. His areas of interest and research cover processes of shifting baselines, humiliation and de-humanization in various social, technical and medial fields; ses@hs-furtwangen.de.

Spindler, Mone, Dr., is a research assistant at the International Centre for Ethics in the Sciences and Humanities (IZEW) at the University of Tübingen. Her research focus is on ageing, sociology of knowledge and body, biopolitics and bioethics; mone.spindler@izew.uni-tuebingen.de.

Turnheim, Bruno, Dr., is postdoctoral Research Associate in the Department of Geography at King's College London (UK). Bruno's research interests include socio-technical transitions, environmental problems and sustainability, path dependence, lock-in, and destabilisation; bruno.turnheim@kcl.ac.uk.

Weinberger, Nora works at the Institute for Technology Assessment and System Analysis (Karlsruhe Institute of Technology – KIT). Her research interests focus on the fields of constructive TA, participatory technology design, demand orientation, ICT in health care, dementia, and sustainability; nora.weinberger@kit.edu.

Wolbring, Gregor is Associate Professor in the Cumming School of Medicine / Dept. of Community Health Sciences at the University of Calgary. His research interests include social evaluation of emerging sciences and technologies, disability studies, ability studies and sustainability studies; gwolbrin@ucalgary.ca.